Seismic Inversion

Theory and Applications

*This book is dedicated to my wife Guo-ling, and
my two children Brian and Claire.*

Seismic Inversion

Theory and Applications

Yanghua Wang

Professor of Geophysics
Imperial College London, UK

WILEY Blackwell

Library of Congress Cataloging-in-Publication Data

Names: Wang, Yanghua.
Title: Seismic inversion : theory and applications / by Yanghua Wang,
   professor of geophysics, Imperial College London, UK.
Description: Malden, MA : Wiley-Blackwell Publishing, Ltd., 2016. | Includes
   bibliographical references and index.
Identifiers: LCCN 2016024723 (print) | LCCN 2016025497 (ebook) | ISBN
   9781119257981 (cloth) | ISBN 9781119258049 (pdf) | ISBN 9781119258025
   (epub)
Subjects: LCSH: Seismic traveltime inversion. | Seismic reflection
   method–Deconvolution. | Seismic tomography. | Seismology–Mathematics.
Classification: LCC QE539.2.S43 W37 2016 (print) | LCC QE539.2.S43 (ebook) |
   DDC 551.22028/7–dc23
LC record available at https://lccn.loc.gov/2016024723

1 2017

Contents

Preface

Seismic inversion aims to reconstruct an Earth subsurface model based on seismic measurements. Such a subsurface model is quantitatively represented by spatially variable physical parameters, and is extracted from seismic data by solving an inverse problem. For seismic inversion, we need to resolve at least three fundamental issues simultaneously: (1) non-linearity, because the solving procedure is dependent upon the solution, that is, seismic wave propagation involved in the inversion is a function of the current model estimate, (2) non-uniqueness due to data incompleteness, and (3) instability, as a small amount of data errors may cause huge perturbations in the model estimate. The last two complicated issues are due to the inverse problem being ill-posed mathematically.

This book introduces the basic theory and solutions of the inverse problems, in correspondence to the above three issues related to seismic inversion. Practically, we must understand the following *how-to*'s: to solve a nonlinear problem by iterative linearisation, to solve an underdetermined problem with model constraints, and to solve an ill-posed problem by regularisations. This book also introduces some applications with which to extract meaningful information from seismic data for reservoir characterisation, in order to stimulate readers' interest for pursuing advanced research in seismic inversion.

This textbook is based on lecture notes of *Seismic Inversion and Quantitative Analysis*, which have been presented to master and doctoral students in geophysics at Imperial College London. The syllabuses are

1) Linear inverse problem
2) Matrix analysis
3) Least-squares method
4) Iterative method
5) Quadratic minimisation
6) Steepest descent method
7) Conjugate gradient method
8) Subspace gradient method
9) Eigenvalues and eigenvectors
10) Singular value analysis
11) Generalised inverse by SVD
12) Maximum entropy method
13) Maximum likelihood method

14) The Cauchy inversion method
15) General L_p norm method
16) Localised average solution
17) Wavelet estimation
18) Reflectivity inversion
19) Ray-impedance inversion
20) Traveltime tomography
21) Waveform tomography for velocity
22) Waveform tomography for impedance

Many of these syllabuses are named as mathematical terminologies. The focus of this book will be their physical meanings. This book is divided into two parts. The first part, consisting of seven chapters, is the fundamentals of linear inverse problems. The second part is methodologies of seismic inversion. The essence of seismic inversion is regularisation. Regularisation can be defined as a model constraint, used additively in an objective function of the inverse problem. Regularisation can also be an action applied directly to the geophysical operator.

Building on the basic theory of linear inverse problems, the methodologies of seismic inversion are explained in detail, including ray-impedance inversion and waveform tomography etc. The application methodologies are categorised into convolutional and wave-equation based groups. This systematic presentation simplifies the subject and enables an in-depth understanding of seismic inversion.

This book summarises the author's extensive experience in both industry and academia and includes innovative techniques not previously published. Conventionally, the convolutional model is used for seismic reflectivity and impedance inversion, and wave equation-based waveform tomography, or full waveform inversion, is inverting for velocity variation. This book presents for the first time the use of the wave equation-based inversion method for the reconstruction of subsurface impedance images.

This book provides a practical guide to reservoir geophysicists who are attempting quantitative reservoir characterisation based on seismic data. Philosophically, the seismic inverse problem allows for a range of possible solutions, but the techniques described herein enable geophysicists to exclude models that cannot satisfy the available data. This book deals with the engineering aspects of the inverse problem, for understanding the mathematical tools and in turn to generate geophysically meaningful solutions.

Yanghua Wang
21 February 2016

CHAPTER 1

Basics of seismic inversion

Seismic inversion attempts to extract spatially variable physical parameters from measured seismic data. These physical parameters may be representative of the Earth's subsurface media, and have physical and geological meanings, and thus seismic inversion is a quantitative interpretation of seismic measurement. The inversion procedure is generally nonlinear, as the entire inversion engine to solve the inverse problem, at least partly, depends upon the solution. In practice, the inverse problem is often linearised, and the final nonlinear solution is obtained through the iterative application of linearised solvers. Therefore, this book will focus on the linear inverse problem.

1.1 The linear inverse problem

Linear seismic inversion may include at least three basic steps such as the following:

1) Setting up an objective function, which describes how well a model estimate represents the seismic observation and meets our human expectation;

2) Optimising the objective function based on a minimal variation principle, which leads to a linear system of equations, if the objective function is defined as a quadratic function;

3) Solving this linear system, to obtain a quantitative solution.

Data fitting is a principal part of the objective function. Seismic inversion

---

Seismic Inversion: Theory and Applications, First Edition. Yanghua Wang.
© 2017 John Wiley & Sons, Ltd. Published 2017 by John Wiley & Sons, Ltd.

uses forward modelling to generate synthetic seismic data that will match the observed seismic data. Forward modelling can be presented in a linear form:

$$\mathbf{Gm} = \mathbf{d},\tag{1.1}$$

where $\mathbf{G}$ is a geophysical operator (a matrix), $\mathbf{m}$ is the 'model' vector, and $\mathbf{d}$ is the 'data' vector. Both vectors $\mathbf{m}$ and $\mathbf{d}$ are defined in the Hilbert space in which the structure (the length and angle) of an inner product of vectors can be measured. Row vectors and column vectors of matrix $\mathbf{G}$ are also defined in the Hilbert space.

For example, for a two-dimensional (2-D) velocity model defined in the $x - z$ domain, we cannot straightforwardly include a fracture or a fault in the model parameterisation. Instead, we shall use an equivalent velocity model, which takes into account the effect of the fracture or the fault, so that the model vector $\mathbf{m}$ is in the Hilbert space and can be involved in any inner product in seismic inversion.

Then, we can define the data-fitting objective function as

$$\phi(\mathbf{m}) = \|\,\tilde{\mathbf{d}} - \mathbf{Gm}\,\|^2,\tag{1.2}$$

where $\tilde{\mathbf{d}}$ is the observed data vector, and $\|\mathbf{r}\|^2 = \mathbf{r}^\mathrm{T}\mathbf{r} = (\mathbf{r},\mathbf{r}) = \sum_i r_i^2$ is the inner product of a single vector $\mathbf{r}$. The symbol $\|\cdot\|$ represents the L_2 norm of a vector.

The optimisation working on the objective function is not necessarily a minimisation only. Depending on the set-up of the objective function, it can also be maximisation. For example, minimising the data misfit is equivalent to the maximisation of the probability. Minimal variation $\partial\phi/\partial\mathbf{m} = \mathbf{0}$, where $\mathbf{0}$ is a null vector, can find either the minimal or maximal extremer of an objective function.

In the objective function of Equation 1.2, $\|\,\tilde{\mathbf{d}} - \mathbf{Gm}\,\|^2$ is the energy of data residuals. The least-squares solution using a minimal variation principle, that is, setting $\partial\phi/\partial\mathbf{m} = \mathbf{0}$, leads to the following linear system:

$$\mathbf{G}^\mathrm{T}\mathbf{Gm} = \mathbf{G}^\mathrm{T}\tilde{\mathbf{d}},\tag{1.3}$$

where $\mathbf{G}^\mathrm{T}$ is the transpose of the rectangular matrix $\mathbf{G}$, and $\mathbf{G}^\mathrm{T}\mathbf{G}$ is a square matrix. A simplified version of Equation 1.3 is

$$\mathbf{Gm} = \tilde{\mathbf{d}}.\tag{1.4}$$

Generally speaking, the inverse problem corresponds to calculating the inverse of the rectangular matrix $\mathbf{G}$ in Equation 1.4. However, this matrix inverse cannot be calculated directly. The problem ultimately corresponds to the least-squares solution of Equation 1.3, which leads to calculating the inverse of the square matrix $\mathbf{G}^{\mathrm{T}}\mathbf{G}$.

In practice, the matrix inverse does not always exist. It means that either the operator $\mathbf{G}$ or $\mathbf{G}^{\mathrm{T}}\mathbf{G}$ are singular. Any modification to the operator is called regularisation, which makes the inverse mapping, from the data space to the model space, happen in a stable and unique way.

For solving the linear system with a large-sized matrix, in order to avoid the direct calculation of the matrix inverse, an iterative method can be used. Each iteration can also be treated as a linear inverse problem, in which the objective function is defined by an error function, and the solution estimate is updated along the (negative) gradient direction. This is called a gradient-based method.

1.2 Data, model and mapping

Let us compare two linear systems presented in Equations 1.1 and 1.4, respectively. Equation 1.1 is a direct problem:

$$(\text{model } M) \Rightarrow [\text{direct mapping } G] \Rightarrow \text{data } D. \qquad (1.5)$$

Given an Earth model M, defined as a set of Earth parameters, and a mapping operator G, the object is to find a set of data D containing all possible measurements in the data space.

Even for this direct problem, M and G may not be unique for a practical problem. For instance, the acoustic, the elastic or the viscoelastic wave equations all can be used for the problem of generating synthetic reflection seismograms. The selection is made based on the practical requirements and *a priori* information.

For a correctly defined physical problem, the direct problem is usually well-posed, which means this mathematical model that describes physical phenomena does have a unique solution, and the solution depends continuously on the model. The continuous dependency means that a small variation $\Delta\mathbf{m}$ in the model space M causes a small perturbation $\Delta\mathbf{d}$ in the data space D.

The inverse problem in Equation 1.4 states that, given a data set D and the mapping operator G, to find a model M:

$$\text{model } M \Leftarrow [\text{direct mapping } G]^{-1} \Leftarrow (\text{data } D). \qquad (1.6)$$

The inversion theory aims to guide the study of inverse problems in order to extract all the information contained in data, while controlling artefacts introduced through the inversion. There are two main kinds of study, as follows:

The first kind is the exact study with perfect data, that is, the study of the existence and uniqueness of solutions and constructing the exact inverse mapping operator. This is a beautiful exercise for classical mathematical analysis, by knowing the statement that the inverse problem is well-posed and readily solved if G is bicontinuous and bijection occurred between spaces M and D. Bicontinuous means that a continuous function also has a continuous inverse function. Bijection means that, for every $\mathbf{d}$ in D, there is exactly one $\mathbf{m}$ in M such that $G(\mathbf{m}) = \mathbf{d}$, and vice versa. However, this inverse problem is not of much interest to applied scientists and engineers.

The other kind is the study of the definitions of generalised solutions and methods for inexact and incomplete data. For the study of this kind of inverse problems, which geophysicists are interested in, the following three remarks should be noted:

1) The inverse mapping operator G^{-1} may not exist. In this case, one should check the definition of the Earth parameters, considering equivalent mapping and *a priori* information.

2) The solution of an inverse problem is often not unique. Hence, the term *solution estimate* instead of solution should be used for the generalised inversion.

3) Different from a direct problem, an inverse problem is usually ill-posed, that is, a small variation of the data leads to uncontrollable perturbation in the solution (Hadamard, 1902).

Mathematically non-continuous mapping operator G^{-1} causes the problem to be ill-posed. For the linear case presented in the previous section, $G(\mathbf{m}) = \mathbf{Gm}$, the study will be on singularity and condition number of $\mathbf{G}$. Non-singularity suggests the existence of the matrix inverse, $\mathbf{G}^{-1}$, and a low condition number indicates that $\mathbf{Gm} = \tilde{\mathbf{d}}$ is a well-posed problem.

1.3 General solutions

Let $\mathbf{d} = G(\mathbf{m})$ be a predicted data set, and $\tilde{\mathbf{d}}$ be an observed data set; a solution $\mathbf{m}$ may be estimated by minimising the distance between $\tilde{\mathbf{d}}$ and $\mathbf{d}$. This type of solution is called a quasi-solution for $\mathbf{m}$.

A quadratic distance is frequently employed in seismic inversion. The quadratic distance between two vectors is measured as

$$\mathrm{dis}(\mathbf{r}_1, \mathbf{r}_2) = \| \mathbf{r}_1 - \mathbf{r}_2 \|, \qquad (1.7)$$

where $\mathbf{r}_1$ and $\mathbf{r}_2$ are two vectors in the same space. The objective function of Equation 1.2 provides a quasi-solution, as it is related to errors in the observed data set, $\tilde{\mathbf{d}} - G(\mathbf{m}) \neq \mathbf{0}$.

As an observed data set is not only inexact, but incomplete as well, an approximate solution, beside the quasi-solution, is also needed. An approximate solution is obtained by minimising the combination of the data-fitting quality criterion and the model choice criterion. The objective function is defined as

$$\phi(\mathbf{m}) = \mathrm{dis}_\mathrm{D}(\tilde{\mathbf{d}}, G(\mathbf{m})) + \mu \mathrm{dis}_\mathrm{M}(\mathbf{m}, \mathbf{m}_\mathrm{ref}), \qquad (1.8)$$

where $\mathrm{dis}_\mathrm{D}(\tilde{\mathbf{d}}, G(\mathbf{m}))$ is the distance defined in the data space D, $\mathrm{dis}_\mathrm{M}(\mathbf{m}, \mathbf{m}_\mathrm{ref})$ is the distance defined in the model space M, and μ is a trade-off parameter balancing the contribution of two criteria in the objective function.

In seismic inverse problems, $\mathbf{m}_\mathrm{ref}$ can be an expected solution. For instance, in the linear case $G(\mathbf{m}) = \mathbf{Gm}$, an objective function can be defined as

$$\phi(\mathbf{m}) = \| \tilde{\mathbf{d}} - \mathbf{Gm} \|^2 + \mu \| \mathbf{m} - \mathbf{m}_\mathrm{ref} \|^2 . \qquad (1.9)$$

The quadratic distance defined by the L_2 norm is a special case of distance measurement, although it is often used in seismic inversion. The distance can be measured in different ways, such as a weighted quadratic distance, frequently employed in seismic inversion.

The objective function in Equations 1.8 or 1.9 is a constrained inverse problem, in which $\mathrm{dis}_\mathrm{M}(\mathbf{m}, \mathbf{m}_\mathrm{ref})$ is a typical model constraint. Different forms of model constraints can be used. Any constraints in the objective function are regularisation working on the geophysical mapping operator, as shown in the following section.

1.4 Regularisation

Regularisation means to suppress singularities that make the problems ill-posed and will cause difficulties in computation. The approximate solution mentioned above is just a practical way to consider the inexact and incomplete data. Regularisation considers the properties of the mapping operator G from the mathematical viewpoint: Whether the numerical instability comes from the singularity, and whether the singular operator can be modified to stabilise the computation.

The stability behaviour means that a small variation in data causes a small perturbation in the solution estimate, and thus depends on the property of the mapping operator. But how strongly is it dependent? Let ε be the vector of the data errors, and $\Delta \mathbf{m}$ the perturbation in the model solution caused by the errors. There are three types of dependences:

1) Linear: $\| \Delta \mathbf{m} \| = \alpha \| \varepsilon \|$;

2) Power law: $\| \Delta \mathbf{m} \| \leq A \| \varepsilon \|^{\alpha}$, $0 < \alpha \leq 1$ and constant A;

3) Logarithmic: $\| \Delta \mathbf{m} \| \propto \left(\ln \dfrac{1}{\| \varepsilon \|} \right)^{-\alpha} = (-\ln \| \varepsilon \|)^{-\alpha}$.

With a linear dependency, it is a well-posed problem. For logarithmic dependency, it is an ill-posed problem. In order to have a stable inversion, at least an operator of power-law dependency, with the exponent α less than 1, should be employed. Unfortunately, the inverse operators in geophysical problems are usually ill-posed with a logarithmic dependency, and thus need to be regularised.

Figure 1.1 displays the dependence of model perturbation $\| \Delta \mathbf{m} \|$ on the data error $\| \varepsilon \|$, with the three relationships: linear (solid curves), $\| \Delta \mathbf{m} \| = \alpha \| \varepsilon \|$; power law (dotted curves), $\| \Delta \mathbf{m} \| = \| \varepsilon \|^{\alpha}$, $0 < \alpha \leq 1$; and logarithmic (dashed curves), $\| \Delta \mathbf{m} \| = (-\ln \| \varepsilon \|)^{-\alpha}$, in which $\| \varepsilon \|$ can be treated as a pre-normalised data error (with the maximum probable error of 1). The three panels (left to right) are cases with $\alpha = 0.3$, 0.6, 0.9, respectively.

Related to this stability issue in the inverse problem, there is a property of the operator, called the condition number. It is defined by the maximum value of the ratio of the relative errors in the model solution to the relative error in the data. If the condition number is small then the error in $\mathbf{m}$ will not be much bigger than the error in $\tilde{\mathbf{d}}$. On the other hand, if the

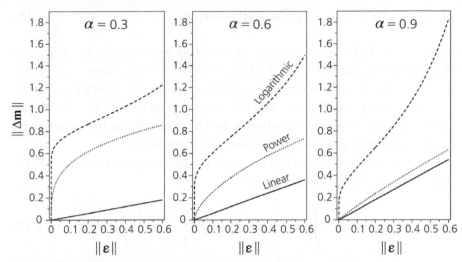

Figure 1.1 The dependence of model perturbation $\| \Delta \mathbf{m} \|$ on the data errors $\| \boldsymbol{\varepsilon} \|$. There are three types of dependence: linear (solid curves), power (dotted curves) and logarithmic (dashed curves). The three panels (left to right) are cases with $\alpha = 0.3, \ 0.6, \ 0.9,$ respectively.

condition number is large, even a small error in data may cause a large error in the model solution. A problem with a low condition number is said to be well-conditioned, whereas a problem with a high condition number is said to be ill-conditioned.

In order to stabilise the inverse problem, by reducing the condition number of the operator, regularisation can be realised as model constraints added to the objective function. It first defines a stabilising function $R(\mathbf{m})$, which satisfies $R(\mathbf{m}) \leq E$ for any real number E, and then incorporates $R(\mathbf{m})$ into the objective function, as

$$\phi(\mathbf{m}) = Q(\mathbf{m}) + \mu R(\mathbf{m}), \tag{1.10}$$

where $Q(\mathbf{m})$ is the data fit quality criterion, and $R(\mathbf{m})$ stands for the model regularisation term.

To understand this stabilisation, let us see an example objective function,

$$\phi(\mathbf{m}) = \| \tilde{\mathbf{d}} - \mathbf{Gm} \|^2 + \mu \| \mathbf{m} \|^2. \tag{1.11}$$

Compared to the objective function in Equation 1.9, $\mathbf{m}_{\mathrm{ref}} = \mathbf{0}$ is set here. Minimisation by setting $\partial \phi / \partial \mathbf{m} = \mathbf{0}$ yields the following equation:

$$[\mathbf{G}^{\mathrm{T}}\mathbf{G} + \mu\mathbf{I}]\,\mathbf{m} = \mathbf{G}^{\mathrm{T}}\tilde{\mathbf{d}}\,. \tag{1.12}$$

If the matrix $\mathbf{G}^{\mathrm{T}}\mathbf{G}$ was singular, the modified operator $[\mathbf{G}^{\mathrm{T}}\mathbf{G} + \mu\mathbf{I}]$ is no longer singular, and the solution of Equation 1.12 exists. Hence, μ is also called the stabilisation factor. The solution estimate $\mathbf{m}$ is unique, as well as continuously dependent on the averaged data, $\mathbf{G}^{\mathrm{T}}\tilde{\mathbf{d}}$. Therefore, constraining the objective function is in fact regularising the geophysical mapping operator, so as to stabilise the inverse problem.

Tikhonov regularisation (Tikhonov, 1935; Tikhonov and Arsenin, 1977; Tikhonov *et al.*, 1995) is expressed as

$$R(\mathbf{m}) = \int_{r_a}^{r_b} \left(\mu_1(r)\,\|\,\mathbf{m}(r)\,\|^2 + \mu_2(r) \left\| \frac{\partial \mathbf{m}(r)}{\partial r} \right\|^2 \right) \mathrm{d}r\,, \tag{1.13}$$

where r is the spatial position, and $\mu_1(r)$ and $\mu_2(r)$ are positive weighting functions, defined within the range $[r_a,\ r_b]$.

Moreover, regularisation can also be applied directly to the geophysical operation, for depressing any singularity. Let us see a simple example, differentiating a continuous function. Assume that $f(r)$ is a real continuous function, but its derivative might not exist. To subjugate this singularity, regularisation can be achieved by convolving $f(r)$ with a continuous and differentiable function $h(r)$,

$$\tilde{f}(r) = f(r) * h(r)\,. \tag{1.14}$$

This processed function $\tilde{f}(r)$ is differentiable without singularities.

In order to make $\tilde{f}(r)$ a good approximation for $f(r)$, the following conditions should be satisfied:

1) Finite range of h: $h(r) \equiv 0$ for r outside a small range;
2) Unimodular:

$$\int_{-\infty}^{\infty} h(r)\,\mathrm{d}r = 1\,; \tag{1.15}$$

3) Approximation:

$$\frac{1}{r_b - r_a} \int_{r_a}^{r_b} |\,\tilde{f}(r) - f(r)\,|\,\mathrm{d}r < \varepsilon\,. \tag{1.16}$$

The first condition means a localised regularisation, the second condition requires that this process does not change the power of the original function, and the last condition, of course, requires the approximation being sufficiently close to the original function. These three conditions are the basic requirement of a regularisation, if the regularisation is directly applied to the geophysical operator.

A demonstration is shown in Figure 1.2. The $f(r)$ function is defined as

$$f(r) = \begin{cases} 2, & r < r_1, \\ \dfrac{r_1 - r - 2r_2}{r_1 - r_2}, & r_1 \le r \le r_2, \\ 1, & r > r_2. \end{cases} \qquad (1.17)$$

This function is continuous, but not differentiable, since its first-order derivative has two singular points at r_1 and r_2. A filter is designed by a Gaussian function,

$$h(r) = \frac{1}{\sigma\sqrt{2\pi}} \exp\left(-\frac{r^2}{2\sigma^2}\right), \qquad (1.18)$$

where σ is the standard deviation. Convolution produces a smooth function $\tilde{f}(r)$.

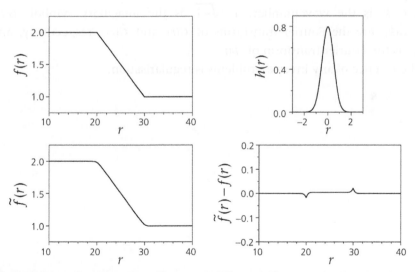

Figure 1.2 A function $f(r)$, that is not differentiable, convolved with a function $h(r)$ produces a differentiable function $\tilde{f}(r)$. The latter is differentiable without singularities, and the difference $|\tilde{f}(r) - f(r)|$ is sufficiently small.

We can verify the numerical example of Figure 1.2 against the three conditions:

1) Localisation of $h(r)$ depends upon the parameter σ. In this display, $\sigma = 0.5$.

2) The filter $h(r)$ is unimodular, because $\sum_i h_i \Delta r = 1$, where Δr is the sampling rate, and $h_i = h(i\Delta r)$.

3) The difference $|\tilde{f}(r) - f(r)|$ is sufficiently small, as the total difference $\varepsilon \leq 0.00126$.

In this simple example, the operator G is the first-order differential,

$$G(r) = \frac{\mathrm{d}}{\mathrm{d}r}. \qquad (1.19)$$

After regularisation, the operator becomes

$$\tilde{G}(r) = \frac{\mathrm{d}}{\mathrm{d}r} h(r) *. \qquad (1.20)$$

These two operators can be understood in the Fourier transform domain as

$$G(k) = -\mathrm{i}k, \qquad \tilde{G}(k) = -\mathrm{i}kH(k), \qquad (1.21)$$

where k is the wavenumber, $\mathrm{i} = \sqrt{-1}$ is the imaginary symbol, $G(k)$ and $\tilde{G}(k)$ are the Fourier transforms of $G(r)$ and $\tilde{G}(r)$, respectively, and $H(k)$ is the Fourier transform of $h(r)$.

The essence of any inverse problems is regularisation.

CHAPTER 2

Linear systems for inversion

The basic inverse problem, as presented in Equation 1.4, is to solve a linear system of equations, $\mathbf{Gm} = \tilde{\mathbf{d}}$, where $\tilde{\mathbf{d}}$ is a vector of data measurements, $\mathbf{m}$ is the model, and $\mathbf{G}$ is the geophysical mapping operator. Many inverse problems have the solution of this linear system as one of their steps.

The geophysical operator $\mathbf{G}$ maps from the model space M to the data space D. Hence, the first step towards inversion is the direct problem:

$$\mathbf{Gm} = \mathbf{d}. \tag{2.1}$$

That is, to construct a linear equation system that relates the geophysical data with the Earth model, $G(\mathbf{m}) = \mathbf{Gm}$. This chapter presents a governing equation that is suitable to different problems in geophysics. The governing equation and its applications to various inverse problems have a solution that can be written as Equation 2.1 in a discrete form.

Therefore, the linear system of Equation 2.1 is a common ground for the inverse problems. Once this common ground is established, the remaining issue will be to find out whether a unique solution exists and whether the problem is stable, even if there are sufficient data that are not contaminated by errors.

2.1 The governing equation and its solution

Different geophysical methods have different partial differential equations, such as the elastic wave equation for seismic exploration, Maxwell's equation for electromagnetic exploration, Poisson's equation for gravity

---

Seismic Inversion: Theory and Applications, First Edition. Yanghua Wang.
© 2017 John Wiley & Sons, Ltd. Published 2017 by John Wiley & Sons, Ltd.

and magnetic exploration. A governing equation that can be used in almost all geophysical methods is

$$\left(\nabla^2 - p\frac{\partial^2}{\partial t^2} - q\frac{\partial}{\partial t}\right)u(\mathbf{r},t) = -\delta(\mathbf{r}-\mathbf{r}_s)s(t)\,, \qquad (2.2)$$

where $u(\mathbf{r},t)$ denotes any kind of geophysical fields, such as an acoustic wavefield, electromagnetic fields, and so on, $\mathbf{r}$ denotes a point in the space, t the time, and $s(t)$ is the corresponding source function at position $\mathbf{r}=\mathbf{r}_s$. This is a typical damping wave equation (Yang, 1997). Table 2.1 lists the physical meaning of the parameters:

  v is seismic wave velocity,
  α is seismic wave attenuation coefficient,
  μ is magnetic permeability,
  ε is dielectric constant,
  σ is electrical conductivity, and
  κ is thermal diffusivity.

Table 2.1 The physical meaning of parameters in the governing equation

Methods	u	p	q
seismic reflection	pressure or displacement	$1/v^2$	$2\alpha/v$
transient EM	magnetic components	$\mu\varepsilon$	$\mu\sigma$
georadar	electrical components	$\mu\varepsilon$	0
low-frequency EM	electromagnetic fields	0	$\mu\sigma$
geothermal exploration	geothermal field	0	$1/\kappa$
electrical, magnetic and gravity	potential fields	0	0

Let us consider a reflection configuration usually used for seismic exploration with a shot-receiver distance (the offset) less than the depth of the detected target. Applying Hooke's law and Newton's second law yields the following equation:

$$MV^2u = \rho \frac{\partial^2 u}{\partial t^2}, \tag{2.3}$$

where M is the elastic modulus, ρ is the density, u is the particle displacement, and t is the time. This is the simplest wave equation. Considering spatial variation both in the media properties and the particle displacement, and meanwhile introducing an external force, the wave equation is properly presented as

$$\left(\nabla^2 - \frac{1}{c^2(\mathbf{r})} \frac{\partial^2}{\partial t^2} \right) u(\mathbf{r}, t) = -\delta(\mathbf{r} - \mathbf{r}_s) s(t), \tag{2.4}$$

where $c = \sqrt{M/\rho}$ is the wave velocity, $\mathbf{r}$ is the space position, and $s(t)$ is a source located at point $\mathbf{r}_s$. This wave equation has the same form as the governing equation with $p = 1/c^2$ and $q = 0$.

In the frequency domain, Equation 2.4 becomes

$$\left(\nabla^2 + \frac{\omega^2}{c^2(\mathbf{r})} \right) u(\mathbf{r}, \omega) = -\delta(\mathbf{r} - \mathbf{r}_s) s(\omega), \tag{2.5}$$

where $u(\mathbf{r}, \omega)$ is the plane wave of constant-frequency ω at the space position $\mathbf{r}$. Considering the underground media to be inhomogeneous and viscoelastic, the velocity c is no longer real, but is complex-valued as $c(i\omega) = \sqrt{M(i\omega)/\rho}$, where $M(i\omega)$ is the complex-valued elastic modulus of frequency ω, then the wave equation can be written as

$$\left(\nabla^2 + \frac{\omega^2}{c^2(\mathbf{r}, i\omega)} \right) u(\mathbf{r}, \omega) = -\delta(\mathbf{r} - \mathbf{r}_s) s(\omega). \tag{2.6}$$

Given the complex velocity $c(i\omega)$, the complex-valued wavenumber may be expressed as

$$\frac{\omega}{c(i\omega)} = \frac{\omega}{v(\omega)} - i\alpha(\omega), \tag{2.7}$$

where $v(\omega) = \sqrt{M_{\mathrm{Re}}(\omega)/\rho}$ is the phase velocity, $\alpha(\omega)$ is the attenuation coefficient, approximated by

$$\alpha(\omega) = \frac{\omega}{2v(\omega)Q(\omega)}, \tag{2.8}$$

and $Q^{-1}(\omega) = M_{\mathrm{Im}}(\omega)/M_{\mathrm{Re}}(\omega)$ is the inverse of the quality factor, measuring the phase delay within the complex modulus $M(i\omega)$.

Inserting the complex wavenumber into Equation 2.6 yields

$$\left(\nabla^2 + \frac{\omega^2}{v^2(\mathbf{r})} - i\omega \frac{2\alpha(\mathbf{r})}{v(\mathbf{r})}\right) u(\mathbf{r}, \omega) = -\delta(\mathbf{r} - \mathbf{r}_s)s(\omega). \tag{2.9}$$

This is a wave equation suitable for seismic exploration where low-frequency waves are usually considered. First, the phase velocity $v(\omega)$, the attenuation coefficient $\alpha(\omega)$, and also the quality factor $Q(\omega)$ can be treated as frequency independent, within the seismic frequency band. Second, a high-order term α^2 is ignored because of $Q^{-2} \ll 1$ for almost all rocks in the low-frequency range. Now, performing an inverse Fourier transform to Equation 2.9 yields a time-domain viscoelastic wave equation as

$$\left(\nabla^2 - \frac{1}{v^2(\mathbf{r})} \frac{\partial^2}{\partial t^2} - \frac{2\alpha(\mathbf{r})}{v(\mathbf{r})} \frac{\partial}{\partial t}\right) u(\mathbf{r}, t) = -\delta(\mathbf{r} - \mathbf{r}_s)s(t). \tag{2.10}$$

This is the same equation as the governing equation (2.2) with parameters $p = 1/v^2$ and $q = 2\alpha/v$.

Different geophysical problems may have the same governing equation, which has a direct solution in the form of a Fredholm integral equation of the first kind:

$$d(t) = \int_\Omega G(\mathbf{r}, t)m(\mathbf{r})\,\mathrm{d}\mathbf{r}. \tag{2.11}$$

This is a direct problem, and can be discretised into a matrix-vector form as in Equation 2.1. An inverse problem is, given the kernel function $G(\mathbf{r}, t)$ and the observation $d(t)$, to find the model $m(\mathbf{r})$.

The Fredholm integral equation is a fundamental equation to be dealt with in geophysical inverse problems. The following three sections show examples that apply the governing equation into seismic inverse problems. They all lead to the same integral equation and, in turn, the same linear system.

2.2 Seismic scattering

Seismic scattering theory is often used in seismic tomography, a seismic inversion method.

According to the concept of wave scattering, the velocity function can be expressed by the sum of two parts: the smoothly varied reference velocity and the velocity disturbance, such as

$$\frac{1}{v^2(\mathbf{r})} = \frac{1 + \beta(\mathbf{r})}{v_0^2(\mathbf{r})},$$ (2.12)

where $v^{-1}(\mathbf{r})$, the reciprocal velocity, is referred to as the slowness, and $\beta(\mathbf{r})$ is the disturbance of square slowness $v_0^{-2}(\mathbf{r})$. The conditions for Equation 2.12 holds are $v(\mathbf{r}) = v_0(\mathbf{r})$ if $|\mathbf{r}| \mapsto \infty$, and $|\beta(\mathbf{r})| \leq 1$. In other words, the disturbance must be localised and small enough.

The corresponding wavefield u should also be divided into two parts:

$$u(\mathbf{r}, t) = u_0(\mathbf{r}, t) + u_{sc}(\mathbf{r}, t),$$ (2.13)

where $u_0(\mathbf{r}, t)$ is the reference wavefield, and $u_{sc}(\mathbf{r}, t)$ is the scattering wavefield, including reflections from the slowness disturbance.

In reflection seismology, if assuming $\alpha(\mathbf{r}) = 0$, Equation 2.10 is reduced to Equation 2.4. Inserting expressions of Equations 2.12 and 2.13 into Equation 2.4, transforming into the frequency domain, and then separating the reference and scattering wavefields, yields the following two equations:

$$\left(\nabla^2 + \frac{\omega^2}{v_0^2(\mathbf{r})} \right) u_0(\mathbf{r}, \mathbf{r}_s, \omega) = -\delta(\mathbf{r} - \mathbf{r}_s) s(\omega),$$ (2.14)

and

$$\left(\nabla^2 + \frac{\omega^2}{v_0^2(\mathbf{r})} \right) u_{sc}(\mathbf{r}, \mathbf{r}_s, \omega) = -\frac{\omega^2}{v_0^2(\mathbf{r})} \beta(\mathbf{r}) u(\mathbf{r}, \mathbf{r}_s, \omega).$$ (2.15)

Assuming $u_{sc} \ll u_0$, Equation 2.15 becomes

$$\left(\nabla^2 + \frac{\omega^2}{v_0^2(\mathbf{r})} \right) u_{sc}(\mathbf{r}, \mathbf{r}_s, \omega) \approx -\frac{\omega^2}{v_0^2(\mathbf{r})} \beta(\mathbf{r}) u_0(\mathbf{r}, \mathbf{r}_s, \omega),$$ (2.16)

where $u_0(\mathbf{r}, \mathbf{r}_s, \omega)$ is referred to as the incident wavefield. This procedure, using the incident field u_0 to replace the total field u as the driving field at each of scatter points, is called the Born approximation in quantum mechanics.

If assuming the source $s(t)$ is an impulse $\delta(t)$ at point $\mathbf{r}_s$, the solution of Equation 2.14 is referred to as Green's function, $g(\mathbf{r}, \mathbf{r}_s, \omega)$, that satisfies

$$\left(\nabla^2 + \frac{\omega^2}{v_0^2(\mathbf{r})} \right) g(\mathbf{r}, \mathbf{r}_s, \omega) = -\delta(\mathbf{r} - \mathbf{r}_s) . \tag{2.17}$$

Then, the actual solution for Equation 2.14 is

$$u_0(\mathbf{r}, \mathbf{r}_s, \omega) = g(\mathbf{r}, \mathbf{r}_s, \omega) s(\mathbf{r}_s, \omega) , \tag{2.18}$$

where $s(\mathbf{r}_s, \omega)$ is the seismic source.

The Born approximation Equation 2.16 is exactly the same as Equation 2.14. Denoting the receiver position by $\mathbf{r}_r$, we can express the solution of Equation (2.16) as

$$u_{sc}(\mathbf{r}_r, \mathbf{r}_s, \omega) \approx \omega^2 \int_{\Omega} \frac{\beta(\mathbf{r})}{v_0^2(\mathbf{r})} g(\mathbf{r}_r, \mathbf{r}, \omega) u_0(\mathbf{r}, \mathbf{r}_s, \omega) \, d\mathbf{r} . \tag{2.19}$$

As Green's function $g(\mathbf{r}, \mathbf{r}_s, \omega)$ is the representative of the incident wavefield $u_0(\mathbf{r}, \mathbf{r}_s, \omega)$, $g(\mathbf{r}_r, \mathbf{r}, \omega)$ is a representative of an imaginary incident wavefield generated by a point source located at the receiver position $\mathbf{r}$, and $g(\mathbf{r}_r, \mathbf{r}, \omega) = g(\mathbf{r}, \mathbf{r}_r, \omega)$.

One may calculate Green's functions numerically by, for example, a finite difference method. But for a special case when the reference velocity $v_0(\mathbf{r})$ is constant, we know that

$$g(\mathbf{r}, \mathbf{r}_s, \omega) = \frac{\exp(ik |\mathbf{r} - \mathbf{r}_s|)}{4\pi |\mathbf{r} - \mathbf{r}_s|} , \tag{2.20}$$

and

$$g(\mathbf{r}_r, \mathbf{r}, \omega) = \frac{\exp(ik |\mathbf{r} - \mathbf{r}_r|)}{4\pi |\mathbf{r} - \mathbf{r}_r|} , \tag{2.21}$$

where $k = \omega / v_0$. Then, Equation 2.19 becomes

$$u_{sc}(\mathbf{r}_r, \mathbf{r}_s, \omega) \approx \left(\frac{k}{4\pi} \right)^2 s(\mathbf{r}_s, \omega) \int_{\Omega} \beta(\mathbf{r}) \frac{\exp[ik(|\mathbf{r} - \mathbf{r}_s| + |\mathbf{r} - \mathbf{r}_r|)]}{|\mathbf{r} - \mathbf{r}_s| \cdot |\mathbf{r} - \mathbf{r}_r|} \, d\mathbf{r} .$$

$$\tag{2.22}$$

This equation is in the form of a Fredholm integral equation of the first kind, and can be used to estimate the slowness disturbance, if the reference velocity is known and Green's functions are calculated (Bleistein, 1984; Bleistein *et al.*, 2000).

2.3 Seismic imaging

The Born approximation can also be used for simulating reflection waves, propagated through a seismic reflectivity model.

Equation 2.16 is a variant of the governing equation. In this equation, we now replace $-\beta(\mathbf{r})/v_0^2(\mathbf{r})$ with $\tilde{R}(\mathbf{r})$, which is a scaled reflectivity. The reflectivity in the acoustic media can be approximated to as

$$R(\mathbf{r}) \approx \frac{v(\mathbf{r}) - v_0(\mathbf{r})}{v(\mathbf{r}) + v_0(\mathbf{r})} \approx \frac{1}{2}\left(1 - \frac{v_0^2(\mathbf{r})}{v^2(\mathbf{r})}\right) = -\frac{1}{2}\beta(\mathbf{r}), \tag{2.23}$$

and the scaled reflectivity is

$$\tilde{R}(\mathbf{r}) = \frac{2}{v_0^2} R(\mathbf{r}) = -\frac{\beta(\mathbf{r})}{v_0^2(\mathbf{r})}. \tag{2.24}$$

Then, we can rewrite Equation 2.16 to

$$\left(\nabla^2 + \frac{\omega^2}{v_0^2(\mathbf{r})}\right) u(\mathbf{r}, \mathbf{r}_s, \omega) = \omega^2 \tilde{R}(\mathbf{r}) \, g(\mathbf{r}, \mathbf{r}_s, \omega) s(\mathbf{r}_s, \omega). \tag{2.25}$$

Here we also replace the incident wavefield $u_0(\mathbf{r}, \mathbf{r}_s, \omega)$ with $g(\mathbf{r}, \mathbf{r}_s, \omega)$ $s(\mathbf{r}_s, \omega)$. A reflection wavefield $u(\mathbf{r}_r, \mathbf{r}_s, \omega)$ recorded at the surface is a part of the entire scattering wavefield generated by the reflectivity model, and can be expressed analytically as

$$u(\mathbf{r}_r, \mathbf{r}_s, \omega) = -\omega^2 \int_\Omega g(\mathbf{r}_r, \mathbf{r}, \omega) \tilde{R}(\mathbf{r}) g(\mathbf{r}, \mathbf{r}_s, \omega) s(\mathbf{r}_s, \omega) \, d\mathbf{r}. \tag{2.26}$$

This is also a Fredholm integral equation of the first kind. The inverse problem which extracts the reflectivity model from seismic reflection data is called seismic imaging.

Practically, instead of the exact inverse operator, an adjoint operator is often used in seismic imaging (Claerbout, 1992). Given a single shot record of the field seismic data $u(\mathbf{r}_r, \mathbf{r}_s, \omega)$, the subsurface reflectivity image is generated by

$$\tilde{R}(\mathbf{r}, \omega) = -\omega^2 \int_{\mathbf{r}_r} \overline{g(\mathbf{r}_r, \mathbf{r}, \omega) g(\mathbf{r}, \mathbf{r}_s, \omega) s(\mathbf{r}_s, \omega)} \, u(\mathbf{r}_r, \mathbf{r}_s, \omega) \, d\mathbf{r}_r, \tag{2.27}$$

where the overbar denotes the complex conjugate. Note that the adjoint operator $\overline{g(\mathbf{r}_r, \mathbf{r}, \omega) g(\mathbf{r}, \mathbf{r}_s, \omega) s(\mathbf{r}_s, \omega)}$ is just an approximation to the actual inverse of the forward modelling operator in Equation 2.26.

Considering the ω^2 factor as two of the first-order derivatives, we may rewrite Equation 2.27 to

$$\tilde{R}(\mathbf{r},\omega) = - \overline{i\omega g(\mathbf{r},\mathbf{r}_s,\omega)s(\mathbf{r}_s,\omega)} \int_{\mathbf{r}_r} i\omega \overline{g(\mathbf{r}_r,\mathbf{r},\omega)}\, u(\mathbf{r}_r,\mathbf{r}_s,\omega)\, d\mathbf{r}_r \,, \qquad (2.28)$$

where $\overline{i\omega g(\mathbf{r},\mathbf{r}_s,\omega)s(\mathbf{r}_s,\omega)}$ is the first-order derivative of a synthetic wavefield, and $i\omega \overline{g(\mathbf{r}_r,\mathbf{r},\omega)}\, u(\mathbf{r}_r,\mathbf{r}_s,\omega)$ is the first-order derivative of a back-propagated wavefield, because the conjugate is a frequency-domain expression for time reversal. This would be an advanced seismic imaging technique, which uses the full wavefield of a shot record, and can be implemented in four steps:

1) Generate the synthetic wavefield by $g(\mathbf{r},\mathbf{r}_s,\omega)s(\mathbf{r}_s,\omega)$;
2) Generate the back-propagation wavefield $\int_{\mathbf{r}_r} \overline{g(\mathbf{r}_r,\mathbf{r},\omega)}\, u(\mathbf{r}_r,\mathbf{r}_s,\omega)\, d\mathbf{r}_r$;
3) Cross-correlate (the first derivatives of) these two wavefields, producing $\tilde{R}(\mathbf{r},\omega)$;
4) Sum over all frequency components, generating an image $\tilde{R}(\mathbf{r},t=0)$.

The final (negative) reflectivity image is simply a stack of images, generated from all individual shot records.

2.4 Seismic downward continuation

The full wavefield imaging technique presented in the previous section is one of seismic migration methods, the imaging techniques to reconstruct a subsurface geophysical model. The most common migration approach is wavefield downward continuation, which projects surface-recorded seismic reflections to each depth level beneath the surface.

First, let us consider an upward continuation problem, as opposed to the wavefield downward continuation. We may present the acoustic wave equation as

$$\begin{cases} \left(\dfrac{\partial^2}{\partial h^2} + \dfrac{\partial^2}{\partial x^2} + \dfrac{\partial^2}{\partial y^2} - \dfrac{1}{v^2}\dfrac{\partial^2}{\partial t^2} \right) u = 0 \,, & 0 < h \le H \,, \\[2mm] u\big|_{h=0} = d(x,y,t) \,, \end{cases} \qquad (2.29)$$

where $u \equiv u(x,y,h,t)$ is the acoustic wavefield at point (x,y,h), v is the velocity, and $d(x,y,t)$ represents seismic data recorded at the Earth's surface $h = 0$. Here, we set up this equation as an initial condition problem, instead of involving a source term $s(x, y, t) \equiv d(x, y, t)$.

Assuming velocity v is a constant, and performing Fourier transform, with respect to x, y and t, we have

$$\begin{cases} \dfrac{\partial^2 u}{\partial h^2} = \left(k_x^2 + k_y^2 - \dfrac{\omega^2}{v^2} \right) u \,, & 0 < h \le H \,, \\[4mm] u\big|_{h=0} = d(k_x, k_y, \omega) \,, \end{cases} \tag{2.30}$$

where k_x and k_y are the wavenumbers along the x and y directions, respectively, and ω is the temporal frequency. Then, we obtain the solution

$$u(k_x, k_y, h, \omega) = m(k_x, k_y, \omega) \exp\left(h \sqrt{k_x^2 + k_y^2 - \dfrac{\omega^2}{v^2}} \right), \tag{2.31}$$

where $m(k_x, k_y, \omega)$ is the initial condition given by

$$m(k_x, k_y, \omega) = \begin{cases} d(k_x, k_y, \omega), & \text{for } k_x^2 + k_y^2 \ge \omega^2 / v^2, \\ 0, & \text{otherwise.} \end{cases} \tag{2.32}$$

Replacing the surface wavefield $d(k_x, k_y, \omega)$ with $m(k_x, k_y, \omega)$ is because the square-root of $k_x^2 + k_y^2 - \omega^2 / v^2$ would be an imaginary value, if $k_x^2 + k_y^2 < \omega^2 / v^2$.

Applying the inverse Fourier transform to Equation 2.31, we may obtain the solution in the temporal-spatial domain as

$$u(x, y, h, t) = \frac{1}{8\pi^3} \int\limits_0^\infty d\omega \iint\limits_{k_x^2 + k_y^2 \ge \omega^2 / v^2} dk_x dk_y \, \exp\left(h \sqrt{k_x^2 + k_y^{2^2} - \frac{\omega^2}{v^2}} \right)$$

$$\times \int\limits_{t \ge 0} dt' \iint\limits_{x, y} dx' dy' m(x', y', t') \exp\left(i \left[k_x(x - x') + k_y(y - y') + \omega(t - t') \right] \right).$$

$$\tag{2.33}$$

This upward continuation procedure is a direct problem to model wavefield $u(x, y, h, t)$ for $h > 0$, given $m(x, y, t)$ at $h = 0$. Here, $h = 0$ is referred to as the plane of source, and any level $h > 0$ as a plane of observation.

Let us now consider an inverse problem of the upward continuation, that is, the downward continuation. If we set *a plane of observation* $h > 0$ as $z = 0$, then *the plane of source* at $h = 0$ becomes $z < 0$. This is equivalent to shifting the depth coordinate upwards, and the Earth's surface is now at $z = 0$. The problem becomes on where we need to find the wavefield $m(x, y, z, t)$ at a depth $z < 0$, given seismic observation $u(x, y, z = 0, t)$ at $z = 0$. Therefore, it turns the downward continuation into the following inverse problem:

$$u(x, y, z = 0, t) = \iiint G_z(x - x', y - y', t - t') \, m(x', y', z, t') \mathrm{d}x' \mathrm{d}y' \mathrm{d}t',$$

$$\text{for } 0 > z \geq -H, \quad (2.34)$$

where G_z is the kernel function defined as

$$G_z(x - x', y - y', t - t') = \frac{1}{8\pi^3} \int_0^\infty \mathrm{d}\omega \iint_{k_x^2 + k_y^2 \geq \omega^2 / v^2} \mathrm{d}k_x \mathrm{d}k_y \, \exp\left(z \sqrt{k_x^2 + k_y^2 - \frac{\omega^2}{v^2}} \right)$$

$$\times \exp\left(\mathrm{i} \left[k_x(x - x') + k_y(y - y') + \omega(t - t') \right] \right).$$

$$(2.35)$$

Once again, Equation 2.34 is a Fredholm integral equation of the first kind. The discretised form of the Fredholm integral equation can be presented in a matrix-vector form as Equation 2.1. Then, an inverse problem involves the calculation of matrix inverse, in which the matrix is the kernel function of Equation 2.35.

Solving the inverse problem, using the linear system of Equation 2.34, produces the seismic wavefield $m(x, y, z, t)$ at variable depth $z < 0$. The final migrated seismic image is merely a display of $m(x, y, z, t = 0)$, where $t = 0$ is known as the imaging condition in seismic migration.

2.5 Seismic data processing

Many procedures and steps in seismic data processing can also be treated as an inverse problem. We list just a few here.

Residual-time tomography. Migration on nonzero offset seismic data generates common-reflection-point (CRP) gathers, on which coherent events might not be flattened yet. An inversion procedure by minimising time-residuals is able to update the velocity model and to flatten the coherent events in CRPs. This is the residual-time tomography.

For traveltime calculation, a ray tracing scheme, following Snell's law or Fermat's traveltime principle, can be used. If a ray path is divided into n

segments with mesh points q_j, for $j = 0, 1, 2, \cdots, N$, the traveltime along the entire ray path $q_j \equiv (x_j, z_j)$ can be estimated numerically by

$$t = \sum_{j=1}^{N} \ell_j s_j ,$$

(2.36)

where ℓ_j is its path length between any two consecutive points q_{j-1} and q_j, and s_j is the slowness, the reciprocal of local velocity v_j.

For multiple rays and traveltimes we can build a system of equations, $\mathbf{Gm} = \mathbf{d}$, where data vector $\mathbf{d}$ is traveltimes, model vector $\mathbf{m}$ is the slowness values, and the geophysical operator $\mathbf{G}$ is constructed by ray tracing. Making a model perturbation to $\mathbf{Gm} = \mathbf{d}$, we obtain the following system:

$$\mathbf{G\Delta m} = \mathbf{\Delta d} .$$

(2.37)

This linear system can be used for residual-time tomography, in which the *data* vector consists of time residuals $\mathbf{\Delta d} = [\Delta t_1, \Delta t_2, \cdots, \Delta t_M]^T$, and the *model* vector consists of model updates $\mathbf{\Delta m} = [\Delta s_1, \Delta s_2, \cdots, \Delta s_N]^T$.

Note that ray paths in $\mathbf{G}$ depend on the velocity model that is updated after solving the system, and thus it is fundamentally a nonlinear problem. In practice, depending upon the complexity, we can decide whether or not to rebuild the operator $\mathbf{G}$ by re-tracing rays through every updated velocity model. If so, we need to go back to ray tracing as in Equation 2.36. There is no such thing as a shortcut to update the operator $\mathbf{G}$.

Interval velocity inversion. In velocity analysis based on the normal moveout (NMO), the velocity picked from the velocity spectrum is approximately equal to the root-mean-square (*rms*) value of the interval velocities. However, if using the relationship between the *rms* velocity and interval velocities, that is, Dix formula, to directly derive the interval velocities, there is often a strong oscillation in the resultant curve of interval velocities along the depth. Sometimes there even exist unphysical values, such as negative velocities. Therefore, we should set it as an inverse problem, and solve the interval velocities with a smooth constraint, the first-order derivative of the solution estimate, presented in Tikhonov regularisation. The objective function is

$$\phi(\mathbf{m}) = \| \mathbf{v}_{rms} - \mathbf{v}(\mathbf{m}) \|^2 + \mu \| \nabla_t \mathbf{m} \|^2 ,$$

(2.38)

where $\mathbf{m}$ is the 'model' vector of interval velocities, $\mathbf{v}_{rms}$ and $\mathbf{v}(\mathbf{m})$ are observed and calculated *rms* velocities, respectively, and ∇_t is the operator of the first-order derivative of the interval velocity with respect to the time, because the velocities are a function of two-way traveltime, t.

Conventionally, the model is defined as a series of interval velocities, $\mathbf{m} = [v_1, v_2, \cdots, v_N]^{\mathrm{T}}$, with pre-set time intervals. In order to have flexible time intervals, we can use a series of Boltzman functions to represent the steps in the interval velocity model (Figure 2.1). For the kth step, the Boltzman function is given by

$$b_k(t) = \frac{v_k - v_{k+1}}{1 + e^{(t-t_k)/c_k}} + v_{k+1}, \qquad (2.39)$$

where v_k and v_{k+1} are the kth and $(k+1)$th interval velocities, respectively, t_k is the time centre of the step, and c_k is a time constant. The centre point of $b_k(t)$ is at $(t_k, \frac{1}{2}(v_k + v_{k+1}))$, and the slope at the centre point is $b_k'(t = t_k) = \frac{1}{4}(v_{k+1} - v_k)/c_k$. It reveals that the time constant c_k determines the slope of this kth step.

For each of Boltzman functions, there are four variables, $\{v_k, v_{k+1}, t_k, c_k\}$. For a pre-set time segment (between two dashed lines in Figure 2.1), a variable t_k means a flexible time interval for defining the interval velocity.

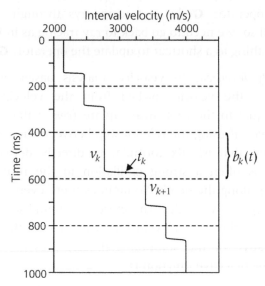

Figure 2.1 A step of the interval velocity model is defined by a Boltzman function, $b_k(t)$, which consists of four parameters $\{v_k, v_{k+1}, t_k, c_k\}$. Within the pre-set time segment (between two dashed lines), inversion produces time t_k for each step. Hence, it has a flexible time interval for defining the interval velocity.

Radon transform. Parabolic Radon transform is often employed to separate seismic multiple reflections from primary reflections.

Suppose a seismic gather is processed with NMO correction, a multiple reflection with the zero-offset time τ is shifted to the time position

$$t(x) = \sqrt{\tau^2 + \frac{x^2}{v_m^2}} - \left(\sqrt{\tau^2 + \frac{x^2}{v_p^2}} - \tau \right), \tag{2.40}$$

where x is the source-receiver offset, v_p is the *rms* velocity of the primary reflection, and v_m is the *rms* velocity of the multiple reflection. The first term on the right-hand side is the hyperbolic curve of the multiple reflection, and the term within the pair of brackets is the NMO quantity, calculated based on the primaries velocity v_p. Making a first-order Taylor expansion to Equation 2.40 produces a parabolic expression

$$t(x) \approx \tau + qx^2, \tag{2.41}$$

where q is the parabolic parameter defined as

$$q = \frac{1}{2\tau} \left(\frac{1}{v_m^2} - \frac{1}{v_p^2} \right). \tag{2.42}$$

Ideally, the parabolic Radon transform can model this seismic gather as a linear combination of parabolic events. However, this so-called transformation does not have rigorously defined forward and inverse transform pairs. In practice, the Radon transform image $u(\tau, q)$ is constructed numerically by solving the following inverse problem:

$$\sum_q u(\tau = t - qx^2, q) = d(t, x), \tag{2.43}$$

where $d(t, x)$ is the input gather to the Radon transform. In the frequency domain, it becomes

$$\sum_q u(\omega, q) \exp(-i\omega q x^2) = d(\omega, x). \tag{2.44}$$

For each individual frequency, this is a linear system (Beylkin, 1987). An inversion method can be employed to find the solution $u(\omega, q)$.

CHAPTER 3

Least-squares solutions

This chapter introduces least-squares solutions to the linear system $\mathbf{Gx} = \mathbf{d}$, where $\mathbf{G}$ is an $M \times N$ rectangular matrix, and $\mathbf{x} \equiv \mathbf{m}$, the model vector. The solutions have a general form of

$$\mathbf{x} = \mathbf{G}^{\dagger}\mathbf{d}, \tag{3.1}$$

where $\mathbf{G}^{\dagger}$ is a pseudo inverse of the matrix $\mathbf{G}$. Thus, matrix computation is fundamentally important for any inverse problem and its least-squares solutions. The least-squares inverse problem to $\mathbf{Gx} = \mathbf{d}$ will lead to a new system

$$\mathbf{G}^{\mathrm{T}}\mathbf{Gx} = \mathbf{G}^{\mathrm{T}}\mathbf{d}, \tag{3.2}$$

where $\mathbf{G}^{\mathrm{T}}$ is the matrix transpose of $\mathbf{G}$. This equation, with a simple pre-multiplication of $\mathbf{G}^{\mathrm{T}}$ to the linear system $\mathbf{Gx} = \mathbf{d}$, is called the normal equation, because $\mathbf{G}^{\mathrm{T}}(\mathbf{d} - \mathbf{Gx}) = \mathbf{0}$, the vector $(\mathbf{d} - \mathbf{Gx})$ being normal to the matrix $\mathbf{G}$.

The linear system in Equation 3.2 can be written as $\mathbf{Ax} = \mathbf{b}$, where $\mathbf{A} = \mathbf{G}^{\mathrm{T}}\mathbf{G}$ is an $N \times N$ square matrix, and $\mathbf{b} = \mathbf{G}^{\mathrm{T}}\mathbf{d}$ is a vector of averaged data. Our discussion will start with the square matrix $\mathbf{A}$, followed by the general rectangular matrix $\mathbf{G}$.

3.1 Determinant and rank

The determinant of a matrix can be understood as a 'volume' in a high dimensional space.

Seismic Inversion: Theory and Applications, First Edition. Yanghua Wang.
© 2017 John Wiley & Sons, Ltd. Published 2017 by John Wiley & Sons, Ltd.

For a 3×3 matrix, for instance, the determinant is

$$\det(\mathbf{A}) = \begin{vmatrix} a_{11} & a_{12} & a_{13} \\ a_{21} & a_{22} & a_{23} \\ a_{31} & a_{32} & a_{33} \end{vmatrix} = a_{11} \begin{vmatrix} a_{22} & a_{23} \\ a_{32} & a_{33} \end{vmatrix} - a_{12} \begin{vmatrix} a_{21} & a_{23} \\ a_{31} & a_{33} \end{vmatrix} + a_{13} \begin{vmatrix} a_{21} & a_{22} \\ a_{31} & a_{32} \end{vmatrix}.$$

$$(3.3)$$

Considering each row of matrix $\mathbf{A}$ as a vector, three row vectors $(\mathbf{r}_1, \mathbf{r}_2, \mathbf{r}_3)$ form a parallelepiped. The volume of this parallelepiped is determined by three row vectors as $|\mathbf{r}_1 \bullet (\mathbf{r}_2 \times \mathbf{r}_3)|$.

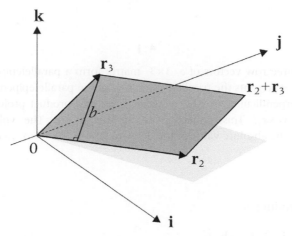

Figure 3.1 Two row vectors $\mathbf{r}_2$ and $\mathbf{r}_3$ form a parallelogram, where b is the height of the parallelogram, and $\mathbf{i}$, $\mathbf{j}$ and $\mathbf{k}$ are the basis vectors.

Two row vectors $\mathbf{r}_2$ and $\mathbf{r}_3$ form a parallelogram (Figure 3.1). The area of this parallelogram is the modular $\| \mathbf{r}_2 \|$ times the height b:

$$\| \mathbf{r}_2 \| b = \| \mathbf{r}_2 \| \| \mathbf{r}_3 \| \sin\theta = \| \mathbf{r}_2 \times \mathbf{r}_3 \|, \qquad (3.4)$$

where θ is the angle between two vectors, and $b = \| \mathbf{r}_3 \| \sin\theta$. Equation 3.4 indicates that the area of the parallelogram is the modular of the cross product $\mathbf{r}_2 \times \mathbf{r}_3$.

While the cross product $\mathbf{r}_2 \times \mathbf{r}_3$, which forms the base of the parallelepiped, has the direction perpendicular to either vectors, the inner product then projects vector $\mathbf{r}_1$ to the vector $\mathbf{r}_2 \times \mathbf{r}_3$. This projection is the height of the parallelepiped (Figure 3.2).

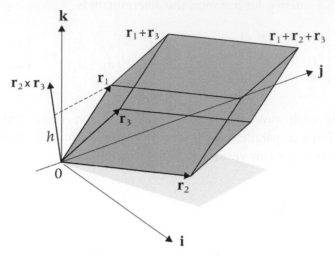

Figure 3.2 Three row vectors of a 3×3 matrix form a parallelepiped. The cross product $\mathbf{r_2} \times \mathbf{r_3}$ represents the parallelogram area of the parallelepiped base and has the direction perpendicular to either vectors. The inner product projects vector $\mathbf{r_1}$ to the vector $\mathbf{r_2} \times \mathbf{r_3}$. This projection is the height h. The volume of this parallelepiped is the absolute value of the determinant of the 3×3 matrix.

The cross product is

$$\mathbf{r_2} \times \mathbf{r_3} = \begin{vmatrix} \mathbf{i} & \mathbf{j} & \mathbf{k} \\ a_{21} & a_{22} & a_{23} \\ a_{31} & a_{32} & a_{33} \end{vmatrix} = \begin{vmatrix} a_{22} & a_{23} \\ a_{32} & a_{33} \end{vmatrix} \mathbf{i} - \begin{vmatrix} a_{21} & a_{23} \\ a_{31} & a_{33} \end{vmatrix} \mathbf{j} + \begin{vmatrix} a_{21} & a_{22} \\ a_{31} & a_{32} \end{vmatrix} \mathbf{k},$$

$$(3.5)$$

where $\mathbf{i}$, $\mathbf{j}$ and $\mathbf{k}$ are the standard basis vectors. Then, the inner product is

$$\mathbf{r_1} \bullet (\mathbf{r_2} \times \mathbf{r_3}) = a_{11} \begin{vmatrix} a_{22} & a_{23} \\ a_{32} & a_{33} \end{vmatrix} - a_{12} \begin{vmatrix} a_{21} & a_{23} \\ a_{31} & a_{33} \end{vmatrix} + a_{13} \begin{vmatrix} a_{21} & a_{21} \\ a_{31} & a_{32} \end{vmatrix}. \qquad (3.6)$$

This is exactly expressed in Equation 3.3, the determinant of the 3×3 matrix. Hence, the absolute value of the determinant is the volume of parallelepiped constructed by three row vectors:

$$|\det(\mathbf{A})| = |\mathbf{r_1} \bullet (\mathbf{r_2} \times \mathbf{r_3})|. \qquad (3.7)$$

For a high-dimensional $N \times N$ matrix, the absolute value of the determinant is the 'volume' of an N dimensional 'parallelepiped' constructed by the N row vectors of the square matrix $\mathbf{A}$.

A cost-effective way to find the determinant of a square matrix $\mathbf{A}$ is via the row transformation. During row operations, replacing the jth row $\mathbf{r}_j$ with $\mathbf{r}_j - c\mathbf{r}_i$, where c is a constant, will not change $\det(\mathbf{A})$. We can transform the square matrix into an upper triangular matrix, for which the product of diagonal elements is the determinant of the matrix.

Let us see an example of a 3×3 matrix,

$$\mathbf{A} = \begin{bmatrix} 2 & 4 & 1 \\ -2 & 3 & 1 \\ -1 & 1 & -3 \end{bmatrix},$$

and perform the elementary row operations, step by step:

$$\begin{matrix} \mathbf{r}_2 + \mathbf{r}_1 \to \\ \mathbf{r}_3 + \frac{1}{2}\mathbf{r}_1 \to \end{matrix} \begin{bmatrix} 2 & 4 & 1 \\ 0 & 7 & 2 \\ 0 & 3 & -\frac{5}{2} \end{bmatrix} \Rightarrow \begin{matrix} \\ \\ \mathbf{r}_3 - \frac{3}{7}\mathbf{r}_2 \to \end{matrix} \begin{bmatrix} 2 & 4 & 1 \\ 0 & 7 & 2 \\ 0 & 0 & -\frac{47}{14} \end{bmatrix}.$$

The product of pivots is the determinant $\det(\mathbf{A}) = -47$.

If any pivot is zero, $\det(\mathbf{A}) = 0$, matrix $\mathbf{A}$ is a singular matrix.

The number of nonzero pivots is the rank of the matrix. The physical meaning of rank is the number of linearly independent row vectors constituting the matrix.

For a rectangular $M \times N$ matrix $\mathbf{G}$, if $\text{rank}(\mathbf{G}) = N \leq M$, it is a full rank matrix. This is just because the square matrix $\mathbf{A} = \mathbf{G}^T\mathbf{G}$ has full rank of N. Let us see an example:

$$\mathbf{G} = \begin{bmatrix} 1 & 2 & 3 \\ 2 & 5 & 7 \\ 3 & 6 & 10 \\ 1 & 2 & 4 \end{bmatrix} \Rightarrow \begin{matrix} \\ \mathbf{r}_2 - 2\mathbf{r}_1 \to \\ \mathbf{r}_3 - 3\mathbf{r}_1 \to \\ \mathbf{r}_4 - \mathbf{r}_1 \to \end{matrix} \begin{bmatrix} 1 & 2 & 3 \\ 0 & 1 & 1 \\ 0 & 0 & 1 \\ 0 & 0 & 1 \end{bmatrix} \Rightarrow \begin{matrix} \\ \\ \\ \mathbf{r}_4 - \mathbf{r}_3 \to \end{matrix} \begin{bmatrix} 1 & 2 & 3 \\ 0 & 1 & 1 \\ 0 & 0 & 1 \\ 0 & 0 & 0 \end{bmatrix}.$$

There are three nonzero pivots, hence $\text{rank}(\mathbf{G}) = 3$.

In the system of linear equations, the rank of a matrix indicates the number of independent equations. If matrix $\mathbf{G}$ is in full rank, the number of independent equations is the same as the number of unknown variables. Then, the system is resolvable. If matrix $\mathbf{G}$ is not in full rank, it is a singular matrix, and its inverse does not exist.

3.2 The inverse of a square matrix

For a square matrix $\mathbf{A}$, if $\det(\mathbf{A}) \neq 0$, it is non-singular, and the matrix inverse $\mathbf{A}^{-1}$ exists:

$$\mathbf{A}^{-1}\mathbf{A} = \mathbf{A}\mathbf{A}^{-1} = \mathbf{I}, \tag{3.8}$$

where $\mathbf{I}$ is the identity matrix. The matrix inverse can be obtained also by the elementary row operation, which is an effective way to solve a system of linear equations.

The system $\mathbf{A}\mathbf{A}^{-1} = \mathbf{I}$ means solving N systems of equations. For example, a 3×3 matrix $\mathbf{A}$ multiplies the first column of $\mathbf{A}^{-1}$ to give the first column of $\mathbf{I}$, multiplies the second column of $\mathbf{A}^{-1}$ to give the second column of $\mathbf{I}$, and multiplies the third column of $\mathbf{A}^{-1}$ to give the third column of $\mathbf{I}$:

$$\mathbf{A}\begin{bmatrix} x_{11} \\ x_{21} \\ x_{31} \end{bmatrix} = \begin{bmatrix} 1 \\ 0 \\ 0 \end{bmatrix}, \qquad \mathbf{A}\begin{bmatrix} x_{12} \\ x_{22} \\ x_{32} \end{bmatrix} = \begin{bmatrix} 0 \\ 1 \\ 0 \end{bmatrix}, \qquad \mathbf{A}\begin{bmatrix} x_{13} \\ x_{23} \\ x_{33} \end{bmatrix} = \begin{bmatrix} 0 \\ 0 \\ 1 \end{bmatrix}. \tag{3.9}$$

The Gauss-Jordan elimination method can be used to solve each of these systems in sequence:

$$\begin{bmatrix} \mathbf{A} & \begin{matrix} 1 \\ 0 \\ 0 \end{matrix} \end{bmatrix}, \qquad \begin{bmatrix} \mathbf{A} & \begin{matrix} 0 \\ 1 \\ 0 \end{matrix} \end{bmatrix}, \qquad \begin{bmatrix} \mathbf{A} & \begin{matrix} 0 \\ 0 \\ 1 \end{matrix} \end{bmatrix}. \tag{3.10}$$

Combining these augmented matrices into one single matrix:

$$\begin{bmatrix} \mathbf{A} & \mathbf{I} \end{bmatrix}, \tag{3.11}$$

then, the Gauss-Jordan elimination method is used on these three systems simultaneously, to solve the matrix inverse. The augmented matrix $[\mathbf{A} \mid \mathbf{I}]$ is row-reduced to $[\mathbf{I} \mid \mathbf{A}^{-1}]$.

For the following 3×3 matrix $\mathbf{A}$,

$$\mathbf{A} = \begin{bmatrix} 2 & 1 & 1 \\ 3 & 2 & 1 \\ 2 & 1 & 2 \end{bmatrix},$$

the augmented matrix $[\mathbf{A} \,|\, \mathbf{I}]$ is

$$\begin{bmatrix} 2 & 1 & 1 & 1 & 0 & 0 \\ 3 & 2 & 1 & 0 & 1 & 0 \\ 2 & 1 & 2 & 0 & 0 & 1 \end{bmatrix}.$$

Performing the elementary row operations step by step, we obtain

$$\begin{bmatrix} 1 & 0 & 0 & 3 & -1 & -1 \\ 0 & 1 & 0 & -4 & 2 & 1 \\ 0 & 0 & 1 & -1 & 0 & 1 \end{bmatrix},$$

in which new pivots are 1. Then, the second half of this six-column matrix is the matrix inverse:

$$\mathbf{A}^{-1} = \begin{bmatrix} 3 & -1 & -1 \\ -4 & 2 & 1 \\ -1 & 0 & 1 \end{bmatrix}.$$

The following three statements are the same thing: The matrix $\mathbf{A}$ is not in full rank; $\det(\mathbf{A}) = 0$; and the matrix is a singular matrix. In this case, there are fewer independent equations than unknowns, and the system is considered to be underdetermined. Extra constraints must be applied, so as to produce a least-squares solution.

If there are more independent equations than unknowns, the system is over-determined. In this case, the least-squares method will also provide an approximate solution, since the system $\mathbf{G}\mathbf{x} = \mathbf{d}$ can be pre-multiplied by a transpose matrix $\mathbf{G}^{\mathrm{T}}$ on both sides: $\mathbf{G}^{\mathrm{T}}\mathbf{G}\mathbf{x} = \mathbf{G}^{\mathrm{T}}\mathbf{d}$.

3.3 LU decomposition and Cholesky factorisation

If a non-singular matrix $\mathbf{G}$ is big, and if we need to solve several different systems with the same matrix $\mathbf{G}$, then we would like to avoid repeating the steps of Gaussian elimination on $\mathbf{G}$ for every different $\mathbf{d}$. This can be accomplished by the LU decomposition.

The main idea of the LU decomposition is to record the steps in the places where the zero is produced by Gaussian elimination. For example, for a 4×3 matrix

$$\mathbf{G} = \begin{bmatrix} a_{11} & a_{12} & a_{13} \\ a_{21} & a_{22} & a_{23} \\ a_{31} & a_{32} & a_{33} \\ a_{41} & a_{42} & a_{43} \end{bmatrix}, \tag{3.12}$$

when using Gaussian elimination to make $\mathbf{G}$ to be an upper triangular matrix, we record steps in a lower triangular matrix:

$$\mathbf{L} = \begin{bmatrix} 1 & 0 & 0 \\ \times & 1 & 0 \\ \times & \times & 1 \\ \times & \times & \times \end{bmatrix}, \tag{3.13}$$

where '×' are places for recording the steps. Note that, in order to make the LU decomposition unique, all elements on the main diagonal of the lower triangular matrix $\mathbf{L}$ are 1.

The first step of Gaussian elimination is to make the element at the second row first column to be zero. The row operation is $\mathbf{r}_2 - \ell_{21}\mathbf{r}_1$, with

$$\ell_{21} = \frac{a_{21}}{a_{11}}. \tag{3.14}$$

To record what we have done, we put the multiplier ℓ_{21} into $\mathbf{L}$ at the place it was used to make a zero. We have the following two matrices:

$$\begin{bmatrix} 1 & 0 & 0 \\ \ell_{21} & 1 & 0 \\ \times & \times & 1 \\ \times & \times & \times \end{bmatrix}, \begin{bmatrix} a_{11} & a_{12} & a_{13} \\ 0 & a_{22} - \ell_{21}a_{12} & a_{23} - \ell_{21}a_{13} \\ a_{31} & a_{32} & a_{33} \\ a_{41} & a_{42} & a_{43} \end{bmatrix}. \tag{3.15}$$

To eliminate the third and fourth rows, first column, the row operations are $\mathbf{r}_3 - \ell_{31}\mathbf{r}_1$ and $\mathbf{r}_4 - \ell_{41}\mathbf{r}_1$, with multipliers defined by

$$\ell_{31} = \frac{a_{31}}{a_{11}}, \qquad \ell_{41} = \frac{a_{41}}{a_{11}}. \tag{3.16}$$

While recording ℓ_{31} and ℓ_{41} in matrix $\mathbf{L}$, we obtain

$$\begin{bmatrix} 1 & 0 & 0 \\ \ell_{21} & 1 & 0 \\ \ell_{31} & \times & 1 \\ \ell_{41} & \times & \times \end{bmatrix}, \begin{bmatrix} a_{11} & a_{12} & a_{13} \\ 0 & a_{22} - \ell_{21}a_{12} & a_{23} - \ell_{21}a_{13} \\ 0 & a_{32} - \ell_{31}a_{12} & a_{33} - \ell_{31}a_{13} \\ 0 & a_{42} - \ell_{41}a_{12} & a_{43} - \ell_{41}a_{13} \end{bmatrix}. \tag{3.17}$$

To eliminate the third and fourth rows, second column, the row operations are $\mathbf{r}_3 - \ell_{32}\mathbf{r}_2$ and $\mathbf{r}_4 - \ell_{42}\mathbf{r}_2$, with multipliers

$$\ell_{32} = \frac{a_{32} - \ell_{31}a_{12}}{a_{22} - \ell_{21}a_{12}}, \qquad \ell_{42} = \frac{a_{42} - \ell_{41}a_{12}}{a_{22} - \ell_{21}a_{12}}. \qquad (3.18)$$

While recording ℓ_{32} and ℓ_{42} in matrix $\mathbf{L}$, we obtain

$$\begin{bmatrix} 1 & 0 & 0 \\ \ell_{21} & 1 & 0 \\ \ell_{31} & \ell_{32} & 1 \\ \ell_{41} & \ell_{42} & \times \end{bmatrix}, \begin{bmatrix} a_{11} & a_{12} & a_{13} \\ 0 & a_{22} - \ell_{21}a_{12} & a_{23} - \ell_{21}a_{13} \\ 0 & 0 & a_{33} - \ell_{31}a_{13} - \ell_{32}(a_{23} - \ell_{21}a_{13}) \\ 0 & 0 & a_{43} - \ell_{41}a_{13} - \ell_{42}(a_{23} - \ell_{21}a_{13}) \end{bmatrix}.$$

$$(3.19)$$

To eliminate the fourth row, third column, the row operation is $\mathbf{r}_4 - \ell_{43}\mathbf{r}_3$ with

$$\ell_{43} = \frac{a_{43} - \ell_{41}a_{13} - \ell_{42}(a_{23} - \ell_{21}a_{13})}{a_{33} - \ell_{31}a_{13} - \ell_{32}(a_{23} - \ell_{21}a_{13})}. \qquad (3.20)$$

This row operation produces the following matrix:

$$\begin{bmatrix} a_{11} & a_{12} & a_{13} \\ 0 & a_{22} - \ell_{21}a_{12} & a_{23} - \ell_{21}a_{13} \\ 0 & 0 & a_{33} - \ell_{31}a_{13} - \ell_{32}(a_{23} - \ell_{21}a_{13}) \\ 0 & 0 & 0 \end{bmatrix}. \qquad (3.21)$$

The final lower triangular 4×3 matrix $\mathbf{L}$ is

$$\mathbf{L} = \begin{bmatrix} 1 & 0 & 0 \\ \ell_{21} & 1 & 0 \\ \ell_{31} & \ell_{32} & 1 \\ \ell_{41} & \ell_{42} & \ell_{43} \end{bmatrix}. \qquad (3.22)$$

Denoting the upper triangular 3×3 matrix (drop off the zero row) as

$$\mathbf{U} = \begin{bmatrix} a_{11} & a_{12} & a_{13} \\ 0 & a_{22} - \ell_{21}a_{12} & a_{23} - \ell_{21}a_{13} \\ 0 & 0 & a_{33} - \ell_{31}a_{13} - \ell_{32}(a_{23} - \ell_{21}a_{13}) \end{bmatrix}, \qquad (3.23)$$

the basic LU decomposition for $\mathbf{G}$ is accomplished:

$$\mathbf{G} = \mathbf{LU}. \qquad (3.24)$$

Let us review Gaussian elimination on the kth column,

$$\begin{bmatrix} a_{1k} \\ \vdots \\ a_{kk} \\ a_{(k+1)k} \\ \vdots \\ a_{Nk} \end{bmatrix} \Rightarrow \begin{bmatrix} a_{1k} \\ \vdots \\ a_{kk} \\ 0 \\ \vdots \\ 0 \end{bmatrix}, \qquad (3.25)$$

whereby we attempt to eliminate elements a_{ik} for $i = k+1, k+2, \cdots, N$. The multipliers are

$$\ell_{ik} = \frac{a_{ik}}{a_{kk}}, \qquad \text{for } i = k+1, \ k+2, \cdots, N. \qquad (3.26)$$

Here, the pivot entry a_{kk} is the denominator, which is required to be nonzero, $a_{kk} \neq 0$. For numerical stability, the pivot entry is required to be distinct from zero, $|a_{kk}| > \varepsilon$, where ε is a small positive value. Therefore, we shall swap rows among $\{k, k+1, \cdots, N\}$ rows, if it is necessary.

This raw-interchanging process is called pivoting, and can be represented as a pivot matrix $\mathbf{P}$. Pre-multiplication $\mathbf{PG}$ means row interchanging of matrix $\mathbf{G}$. For instance, a pivot matrix is represented as

$$\mathbf{P} = \begin{bmatrix} 1 & 0 & 0 & 0 \\ 0 & 0 & 1 & 0 \\ 0 & 1 & 0 & 0 \\ 0 & 0 & 0 & 1 \end{bmatrix}. \qquad (3.27)$$

Matrix multiplication $\mathbf{PG}$ means the second and third rows of $\mathbf{G}$ are switched by pivoting.

The pivot matrix $\mathbf{P}$ is a permutation matrix, as it is a square binary matrix that has exactly one entry 1 in each row and each column and 0s elsewhere. The permutation matrix is an orthogonal matrix, $\mathbf{PP}^{-1} = \mathbf{PP}^{T} = \mathbf{I}$.

If we also include pivoting, then an LU decomposition for $\mathbf{G}$ consists of three matrices $\mathbf{P}$, $\mathbf{L}$ and $\mathbf{U}$ such that

$$\mathbf{PG} = \mathbf{LU}. \qquad (3.28)$$

To use this information to solve $\mathbf{Gx} = \mathbf{d}$, we first pivot both sides by multiplying the pivot matrix:

$$\mathbf{PGx} = \mathbf{Pd} \equiv \mathbf{b}. \tag{3.29}$$

Substituting LU for **PG** we get

$$\mathbf{LUx} = \mathbf{b}. \tag{3.30}$$

Then, we need only to solve a forward substitution problem,

$$\mathbf{Ly} = \mathbf{b}, \tag{3.31}$$

and analogously a back substitution problem,

$$\mathbf{Ux} = \mathbf{y}. \tag{3.32}$$

We can then solve for any other **d** without redoing LU decomposition.

In many applications where linear systems appear, one needs to solve **Gx** = **d** for many different vectors **d**. For example, in frequency-domain waveform tomography (Chapter 12), we solve this equation the first time for a synthetic wavefield, where **d** is the source wavelet. Then, we use data residuals (the difference between the observed and the synthetic seismic data) iteratively as a virtual sources. For each iteration we solve the problem without redoing the LU decomposition step, but just reloading the vector **d**.

The LU decomposition above could become a Cholesky factorisation, if a matrix is square and symmetric, such as the normal matrix $\mathbf{A} = \mathbf{G}^{\mathrm{T}}\mathbf{G}$ in the normal equation (Equation 3.2), which is the least-squares formulation of the inverse problem and will be described in the next section. For an over-determined problem, the symmetric matrix **A** often is positive definite, which means the quadratic form $\mathbf{z}^{\mathrm{T}}\mathbf{Az} > 0$, for all nonzero vectors **z**. Just as we can take a positive square root of a positive number, so can we take a positive 'square root' of a positive definite matrix. This is the Cholesky factorisation:

$$\mathbf{A} = \mathbf{LL}^{\mathrm{T}}, \tag{3.33}$$

where **L** is a lower triangular matrix. Unlike LU decomposition, the elements on the main diagonal of **L** are not necessarily to be '1'. The upper triangular matrix **U** is just the transpose of the lower triangular matrix, $\mathbf{L}^{\mathrm{T}}$.

The Cholesky factorisation is particularly easy. Let us partition the $N \times N$ matrices

$$\mathbf{A} = \begin{bmatrix} \alpha_{11} & \circ \\ \mathbf{a}_{21} & \mathbf{A}_{22} \end{bmatrix}, \qquad \mathbf{L} = \begin{bmatrix} \lambda_{11} & \circ \\ \mathbf{l}_{21} & \mathbf{L}_{22} \end{bmatrix}, \qquad (3.34)$$

where Greek lower case letters (α_{11} and λ_{11}) refer to scalars, bold lower case letters ($\mathbf{a}_{21}$ and $\mathbf{l}_{21}$) refer to (column) vectors, and bold upper case letters ($\mathbf{A}_{22}$ and $\mathbf{L}_{22}$) refer to matrices. The $\circ$ refers to a part of $\mathbf{A}$ that is neither stored nor updated. By substituting these partitioned matrices into $\mathbf{A} = \mathbf{LL}^{\mathrm{T}}$, we have

$$\begin{bmatrix} \alpha_{11} & \circ \\ \mathbf{a}_{21} & \mathbf{A}_{22} \end{bmatrix} = \begin{bmatrix} \lambda_{11} & \circ \\ \mathbf{l}_{21} & \mathbf{L}_{22} \end{bmatrix} \begin{bmatrix} \lambda_{11} & \mathbf{l}_{21}^{\mathrm{T}} \\ \circ & \mathbf{L}_{22}^{\mathrm{T}} \end{bmatrix} = \begin{bmatrix} \lambda_{11}^2 & \circ \\ \lambda_{11} \mathbf{l}_{21} & \mathbf{l}_{21} \mathbf{l}_{21}^{\mathrm{T}} + \mathbf{L}_{22} \mathbf{L}_{22}^{\mathrm{T}} \end{bmatrix}.$$

$$(3.35)$$

Hence, we find that

$$\mathbf{L} = \begin{bmatrix} \lambda_{11} = \sqrt{a_{11}} & \circ \\ \mathbf{l}_{21} = \mathbf{a}_{21} / \lambda_{11} & \mathbf{L}_{22} = \mathrm{Chol}(\mathbf{A}_{22} - \mathbf{l}_{21} \mathbf{l}_{21}^{\mathrm{T}}) \end{bmatrix}, \qquad (3.36)$$

where $\mathrm{Chol}(\mathbf{A}_{22} - \mathbf{l}_{21} \mathbf{l}_{21}^{\mathrm{T}})$ is the Cholesky factorisation of order $N - 1$.

Explicitly, Cholesky factorisation can be implemented in a recurrent form as

$$\lambda_{ii} = \sqrt{a_{ii} - \sum_{k=1}^{i-1} \lambda_{ik}^2}, \qquad (3.37)$$

$$\lambda_{ij} = \frac{1}{\lambda_{ii}} \left(a_{ij} - \sum_{k=1}^{i-1} \lambda_{ik} \lambda_{jk} \right), \qquad (3.38)$$

where i is the row index, and j is the column index.

Cholesky factorisation is more efficient than LU decomposition, and has excellent numerical stability properties. However, if the symmetric matrix $\mathbf{A}$ is positive semidefinite, which means the quadratic form $\mathbf{z}^{\mathrm{T}} \mathbf{A} \mathbf{z} \geq 0$ (nonnegative) for all nonzero vectors $\mathbf{z}$, a Cholesky factorisation exists but the triangular matrix $\mathbf{L}$ possibly has some zero elements on the diagonal. To stabilise the calculation, Cholesky factorisation should be implemented with complete pivoting, which at each stage permutes the largest diagonal element in the active submatrix into the pivot position.

Once the Cholesky factorisation of $\mathbf{A}$ is available, it is straightforward to solve the linear system $\mathbf{A} \mathbf{x} = \mathbf{b}$, through two iterative substitutions:

$$\mathbf{L}\mathbf{y} = \mathbf{b},$$
$$\mathbf{L}^{\mathrm{T}}\mathbf{x} = \mathbf{y}.$$
(3.39)

Using either LU Decomposition or Cholesky Factorisation, the matrix inverse is not required in solving the linear problem.

3.4 Least-squares solutions of linear systems

If, and only if, $\mathbf{G}^{\mathrm{T}}\mathbf{G}$ is non-singular or full rank, the inverse of $\mathbf{G}^{\mathrm{T}}\mathbf{G}$ exists and the normal equation (Equation 3.2), $\mathbf{G}^{\mathrm{T}}\mathbf{G}\mathbf{x} = \mathbf{G}^{\mathrm{T}}\mathbf{d}$, has a unique solution as

$$\mathbf{x} = [\mathbf{G}^{\mathrm{T}}\mathbf{G}]^{-1}\mathbf{G}^{\mathrm{T}}\mathbf{d}.$$
(3.40)

This is a least-squares solution to the system $\mathbf{G}\mathbf{x} = \mathbf{d}$ with a rectangular $M \times N$ matrix $\mathbf{G}$.

To prove it, we simply set up the following L_2 norm function:

$$\phi(\mathbf{x}) = \| \mathbf{d} - \mathbf{G}\mathbf{x} \|^2.$$
(3.41)

Note that the square of L_2 norm is the inner product of the single vector. Minimisation by setting

$$\frac{\partial \phi(\mathbf{x})}{\partial \mathbf{x}} = -2\mathbf{G}^{\mathrm{T}}(\mathbf{d} - \mathbf{G}\mathbf{x}) = 0,$$
(3.42)

leads to the normal equation. Therefore, Equation 3.40 is a least-squares solution derived from a minimum residual energy criterion.

To further explain the meaning of 'least squares', let us see the following example with $N = 3$. In a scientific experiment, we have three pairs of measurements (α_i, β_i), where $i = 1, 2, 3$, and $\alpha_1 \neq \alpha_2 \neq \alpha_3$. We attempt to fit a straight line $y = a_0 + a_1 x$ to these three points (Figure 3.3):

$$a_0 + a_1\alpha_i = \beta_i,$$
(3.43)

for $i = 1, 2, 3$. The error at each point is the difference between the value of y given by the line and the measured value β_i, that is

$$e_i = a_0 + a_1\alpha_i - \beta_i,$$
(3.44)

for $i = 1, 2, 3$. We wish to choose a_0 and a_1 so that the total of these errors is as small as possible. Since the errors e_i may be positive or

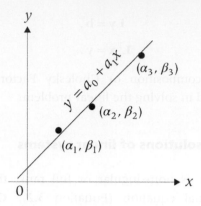

Figure 3.3 Fitting a straight line to three available measurements is a least-squares inverse problem. In this case, the number of data samples is more than the number of unknown variables (a_0, a_1).

negative and we are interested in the net total, we consider the squares of the differences, e_i^2. The least-squares method is to choose a_0 and a_1 so that the sum of the squares of the errors

$$\phi = \sum_{i=1}^{3} (a_0 + a_1\alpha_i - \beta_i)^2 \tag{3.45}$$

is minimised.

For ϕ to be a minimum, each of the partial derivatives $\partial\phi/\partial a_0$ and $\partial\phi/\partial a_1$ must be zero:

$$\frac{\partial\phi}{\partial a_0} = 2\sum_{i=1}^{3}(a_0 + a_1\alpha_i - \beta_i) = 0\,,$$

$$\frac{\partial\phi}{\partial a_1} = 2\sum_{i=1}^{3}\alpha_i(a_0 + a_1\alpha_i - \beta_i) = 0\,. \tag{3.46}$$

Simplifying these two equations gives, respectively,

$$3a_0 + a_1\sum_{i=1}^{3}\alpha_i = \sum_{i=1}^{3}\beta_i\,,$$

$$a_0\sum_{i=1}^{3}\alpha_i + a_1\sum_{i=1}^{3}\alpha_i^2 = \sum_{i=1}^{3}\alpha_i\beta_i\,. \tag{3.47}$$

If Equation 3.46 is written in the form of $\mathbf{Gx} = \mathbf{d}$ with

$$\mathbf{G} = \begin{bmatrix} 1 & \alpha_1 \\ 1 & \alpha_2 \\ 1 & \alpha_3 \end{bmatrix}, \qquad \mathbf{x} = \begin{bmatrix} a_0 \\ a_1 \end{bmatrix}, \qquad \mathbf{d} = \begin{bmatrix} \beta_1 \\ \beta_2 \\ \beta_3 \end{bmatrix}, \qquad (3.48)$$

then it is easy to verify that the pair in Equation 3.47 is exactly the same equation as the normal equation, $\mathbf{G}^T\mathbf{Gx} = \mathbf{G}^T\mathbf{d}$.

When $\alpha_1 \neq \alpha_2 \neq \alpha_3$, $\text{rank}(\mathbf{G}) = 2$, then $\text{rank}(\mathbf{G}^T\mathbf{G}) = 2$ also, and the 2×2 matrix $\mathbf{G}^T\mathbf{G}$ is non-singular. Hence, the inverse of matrix $\mathbf{G}^T\mathbf{G}$ exists and the problem has a unique solution as in Equation 3.40.

However, if the square matrix $\mathbf{A} = \mathbf{G}^T\mathbf{G}$ is singular, a small positive value μ may be added to the diagonal:

$$[\mathbf{G}^T\mathbf{G} + \mu\mathbf{I}]\mathbf{x} = \mathbf{G}^T\mathbf{d}, \qquad (3.49)$$

and the solution can be expressed as

$$\mathbf{x} = [\mathbf{G}^T\mathbf{G} + \mu\mathbf{I}]^{-1}\mathbf{G}^T\mathbf{d}. \qquad (3.50)$$

This is a stabilised least-squares solution, in which μ is a stabilisation factor, for stabilising the matrix inverse calculation.

Equation 3.49 can also be derived from minimising the following regularised objective function:

$$\phi(\mathbf{x}) = \|\mathbf{d} - \mathbf{Gx}\|^2 + \mu\|\mathbf{x}\|^2. \qquad (3.51)$$

Setting $\partial\phi(\mathbf{x})/\partial\mathbf{x} = \mathbf{0}$ will lead to Equation 3.49. Therefore, parameter μ is also a trade-off parameter that balances the residual energy $\|\mathbf{d} - \mathbf{Gx}\|^2$ and the regularisation $\|\mathbf{x}\|^2$.

In general, when the objective function is presented as a weighted sum:

$$\phi(\mathbf{x}) = \|\mathbf{d} - \mathbf{Gx}\|^2 + \mu\|\mathbf{g} - \mathbf{Fx}\|^2, \qquad (3.52)$$

it can also be expressed as an ordinary least-squares objective:

$$\phi(\mathbf{x}) = \left\| \begin{bmatrix} \mathbf{d} \\ \sqrt{\mu}\mathbf{g} \end{bmatrix} - \begin{bmatrix} \mathbf{G} \\ \sqrt{\mu}\mathbf{F} \end{bmatrix} \mathbf{x} \right\|^2 = \|\tilde{\mathbf{d}} - \tilde{\mathbf{G}}\mathbf{x}\|^2, \qquad (3.53)$$

where

$$\tilde{\mathbf{d}} = \begin{bmatrix} \mathbf{d} \\ \sqrt{\mu}\mathbf{g} \end{bmatrix}, \qquad \tilde{\mathbf{G}} = \begin{bmatrix} \mathbf{G} \\ \sqrt{\mu}\mathbf{F} \end{bmatrix}. \qquad (3.54)$$

Assuming that the augmented $\tilde{\mathbf{G}}$ is in full rank, the solution is

$$\mathbf{x} = [\tilde{\mathbf{G}}^{\mathrm{T}}\tilde{\mathbf{G}}]^{-1}\tilde{\mathbf{G}}^{\mathrm{T}}\tilde{\mathbf{d}}$$
$$= [\mathbf{G}^{\mathrm{T}}\mathbf{G} + \mu\mathbf{F}^{\mathrm{T}}\mathbf{F}]^{-1}(\mathbf{G}^{\mathrm{T}}\mathbf{d} + \mu\mathbf{F}^{\mathrm{T}}\mathbf{g}). \qquad (3.55)$$

This is a unified solution, for example, to the objective functions with $\mathbf{F} = \mathbf{I}$ (Equation 3.51), or the first-order derivative with respect to the model parameters (Equation 2.38).

3.5 Least-squares solution for a nonlinear system

For a nonlinear problem,

$$\phi(\mathbf{x}) = \| \mathbf{d} - G(\mathbf{x}) \|^2 = \| \mathbf{e}(\mathbf{x}) \|^2, \qquad (3.56)$$

we linearise $\mathbf{e}(\mathbf{x})$ by making a first-order Taylor expansion near the current estimate $\mathbf{x}^{(k)}$:

$$\phi(\mathbf{x}) \approx \| \mathbf{e}(\mathbf{x}^{(k)}) + \mathbf{D}^{(k)}(\mathbf{x} - \mathbf{x}^{(k)}) \|^2 = \| \mathbf{D}^{(k)}\mathbf{x} - \mathbf{b}^{(k)} \|^2, \qquad (3.57)$$

where $\mathbf{D}$ is the Jacobian $\mathbf{D} = \partial\mathbf{e}/\partial\mathbf{x}$, k is the iteration number, and $\mathbf{b}^{(k)} = \mathbf{D}^{(k)}\mathbf{x}^{(k)} - \mathbf{e}(\mathbf{x}^{(k)})$. Hence, the least-squares solution of this linearised problem is

$$\mathbf{x}^{(k+1)} = [(\mathbf{D}^{(k)})^{\mathrm{T}}\mathbf{D}^{(k)}]^{-1}(\mathbf{D}^{(k)})^{\mathrm{T}}\mathbf{b}^{(k)}. \qquad (3.58)$$

This solution is repeated until convergence, for which a basic requirement is the initial estimate $\mathbf{x}^{(0)}$ being close to the true solution.

However, the convergence is not guaranteed. We could simply add a regularisation term, as

$$\phi(\mathbf{x}) = \| \mathbf{D}^{(k)}\mathbf{x} - \mathbf{b}^{(k)} \|^2 + \mu \| \mathbf{x} - \mathbf{x}^{(k)} \|^2, \qquad (3.59)$$

so that the solution of the next iteration will not be too far away from the current solution. For a highly nonlinear problem, extra physical constraints can be enforced. For example, in seismic ray tracing, Fermat's minimum time principle can be exploited, so that the inversion at each iteration will minimise the travel time, rather than the error vector $\mathbf{e}$ (Wang, 2014).

3.6 Least-squares solution by QR decomposition

For a full rank $M \times N$ matrix $\mathbf{G}$, rank($\mathbf{G}$) $= N \leq M$, QR decomposition may transfer it to a much simpler form. Then, we can solve the least-squares problem directly and easily for equation $\mathbf{Gx} = \mathbf{d}$.

A suitably chosen orthogonal matrix $\mathbf{Q}$ will triangularise the given matrix:

$$\mathbf{G} = \mathbf{Q}\begin{bmatrix} \mathbf{R} \\ \mathbf{0} \end{bmatrix}, \tag{3.60}$$

where $\mathbf{Q}$ is an $M \times M$ orthogonal matrix with $\mathbf{Q}^{\mathrm{T}}\mathbf{Q} = \mathbf{I}$, and $\mathbf{R}$ is an $N \times N$ upper triangular matrix.

Consider a 2×1 vector $\mathbf{r} = [r_1 \ r_2]^{\mathrm{T}}$, we need to find a 2×2 orthogonal matrix

$$\mathbf{P} = \begin{bmatrix} c & s \\ -s & c \end{bmatrix}, \tag{3.61}$$

so that $\mathbf{P}^{\mathrm{T}}\mathbf{P} = \mathbf{I}$; this leads to $c^2 + s^2 = 1$. Pre-multiplying the orthogonal matrix $\mathbf{P}$ to the vector $\mathbf{r}$ will make one element of $\mathbf{r}$ be 0:

$$\mathbf{Pr} = \begin{bmatrix} c & s \\ -s & c \end{bmatrix}\begin{bmatrix} r_1 \\ r_2 \end{bmatrix} = \begin{bmatrix} \rho \\ 0 \end{bmatrix}. \tag{3.62}$$

Therefore, we solve this linear system of equations, and obtain

$$\rho = \sqrt{r_1^2 + r_2^2}, \qquad c = \frac{r_1}{\rho}, \qquad s = \frac{r_2}{\rho}. \tag{3.63}$$

This solution reveals that the orthogonal matrix is a clockwise rotation matrix with angle θ, whereas $c = \cos\theta$ and $s = \sin\theta$. Every time, applying such a clockwise coordinate rotation process, will make one element of matrix $\mathbf{G}$ be equal to 0.

For speeding up QR decomposition, the Householder transformation (Householder, 1955, 1964) can directly process a column vector, rather than processing one element at a time. For example, for the first three columns:

$$\mathbf{P}_1 \begin{bmatrix} \times \\ \times \\ \times \\ \times \\ \times \\ \times \end{bmatrix} = \begin{bmatrix} \otimes \\ 0 \\ 0 \\ \vdots \\ \vdots \\ 0 \end{bmatrix}, \quad \mathbf{P}_2 \begin{bmatrix} \times \\ \times \\ \times \\ \times \\ \times \\ \times \end{bmatrix} = \begin{bmatrix} \times \\ \otimes \\ 0 \\ \vdots \\ \vdots \\ 0 \end{bmatrix}, \quad \mathbf{P}_3 \begin{bmatrix} \times \\ \times \\ \times \\ \times \\ \times \\ \times \end{bmatrix} = \begin{bmatrix} \times \\ \times \\ \otimes \\ 0 \\ \vdots \\ 0 \end{bmatrix}, \tag{3.64}$$

where $\mathbf{P}_k$ are orthogonal matrices, and $\times$ denotes any nonzero elements in a vector. The main diagonal elements ($\otimes$) are changed during the column transformation, and all elements below the main diagonal are zero-valued. The final upper triangular matrix has only $\frac{1}{2}N(N+1)$ nonzero elements left. The details of the Householder method are summarised in Appendix A.

By denoting an $M \times M$ matrix $\mathbf{Q}^T$ as the multiplication of orthogonal matrices $\mathbf{Q}^T = \mathbf{P}_\ell \cdots \mathbf{P}_2 \mathbf{P}_1$, we have $\mathbf{G} = \mathbf{QR}$. Since $\mathbf{Q}^T$ is formed by the orthogonal matrices $\mathbf{P}_k$, it is also an orthogonal matrix.

After the QR factorisation, we only have to solve the following triangular system:

$$\mathbf{Rx} = \mathbf{Q}^T_{N \times M} \mathbf{d}, \tag{3.65}$$

where $\mathbf{Q}^T_{N \times M}$ consists of the first N rows of $\mathbf{Q}^T$, i.e. the first N columns of $\mathbf{Q}$. This triangular system can be solved easily by back substitution.

The solution of the triangular system in Equation 3.65 may be expressed as

$$\mathbf{x} = \mathbf{R}^{-1} \mathbf{Q}^T_{N \times M} \mathbf{d}. \tag{3.66}$$

We can see that this is also a least-squares solution. The least-squares solution $\mathbf{x} = [\mathbf{G}^T \mathbf{G}]^{-1} \mathbf{G}^T \mathbf{d}$, which is the solution of $\mathbf{Gx} = \mathbf{d}$, becomes

$$\mathbf{x} = \left(\begin{bmatrix} \mathbf{R} \\ \mathbf{0} \end{bmatrix}^T \mathbf{Q}^T \mathbf{Q} \begin{bmatrix} \mathbf{R} \\ \mathbf{0} \end{bmatrix} \right)^{-1} \begin{bmatrix} \mathbf{R} \\ \mathbf{0} \end{bmatrix}^T \mathbf{Q}^T \mathbf{d}$$

$$= [\mathbf{R}^T \mathbf{R}]^{-1} \mathbf{R}^T \mathbf{Q}^T_{N \times M} \mathbf{d} \tag{3.67}$$

$$= \mathbf{R}^{-1} \mathbf{Q}^T_{N \times M} \mathbf{d}.$$

In the least-squares problem for $\mathbf{Gx} = \mathbf{d}$, we try to minimise the error measured by the L_2 norm $\| \mathbf{d} - \mathbf{Gx} \|$. In the least-squares problem for

Equation 3.65, the L_2 norm becomes $\| \mathbf{Q}^T\mathbf{d} - \mathbf{Q}^T\mathbf{Gx} \|$. It is straight-forward to prove that for any orthogonal $\mathbf{Q}$, this L_2 norm is unchanged by transformation of $\mathbf{Q}^T$:

$$\| \mathbf{Q}^T\mathbf{d} - \mathbf{Q}^T\mathbf{Gx} \| = \| \mathbf{d} - \mathbf{Gx} \|. \tag{3.68}$$

However, $\mathbf{Q}^T\mathbf{G}$ has a particularly attractive form, an upper triangular matrix, $\mathbf{Q}^T\mathbf{G} = \mathbf{R}$.

In summary, this chapter presents four types of least-squares methods. Among these four solutions, the first three need to solve the inverse of a square matrix, $[\mathbf{G}^T\mathbf{G}]$, $[\mathbf{G}^T\mathbf{G} + \mu\mathbf{I}]$ or $[\mathbf{G}^T\mathbf{G} + \mu\mathbf{F}^T\mathbf{F}]$. In the fourth method, pre-multiplying an orthogonal matrix $\mathbf{Q}^T$ can transform them to be an upper triangular matrix. Then, the upper triangular system can be solved easily by back-substitution.

The objective function $\phi(\mathbf{x})$	Least-squares solutions $\mathbf{x}$
$\phi(\mathbf{x}) = \| \mathbf{d} - \mathbf{Gx} \|^2$,	$\mathbf{x} = [\mathbf{G}^T\mathbf{G}]^{-1}\mathbf{G}^T\mathbf{d}$;
$\phi(\mathbf{x}) = \| \mathbf{d} - \mathbf{Gx} \|^2 + \mu\| \mathbf{x} \|^2$,	$\mathbf{x} = [\mathbf{G}^T\mathbf{G} + \mu\mathbf{I}]^{-1}\mathbf{G}^T\mathbf{d}$;
$\phi(\mathbf{x}) = \| \mathbf{d} - \mathbf{Gx} \|^2 + \mu\| \mathbf{g} - \mathbf{Fx} \|^2$,	$\mathbf{x} = [\mathbf{G}^T\mathbf{G} + \mu\mathbf{F}^T\mathbf{F}]^{-1}(\mathbf{G}^T\mathbf{d} + \mu\mathbf{F}^T\mathbf{g})$;
$\phi(\mathbf{x}) = \| \mathbf{Q}^T\mathbf{d} - \mathbf{Q}^T\mathbf{Gx} \|^2$,	$\mathbf{x} = [\mathbf{Q}^T\mathbf{G}]^{-1}\mathbf{Q}^T\mathbf{d}$.

Pre-multiplication of an orthogonal matrix represents physically the processing of rotation and/or reflection. The next chapter will show further matrix manipulation, by pre- and post-multiplying orthogonal matrices, to make a diagonal matrix.

CHAPTER 4

Singular value analysis

This chapter presents the concept of 'singular values' for a rectangular matrix, the algorithm of singular value decomposition (SVD), and SVD-based solutions to linear inverse problems.

A pseudo inverse of the $M \times N$ rectangular matrix $\mathbf{G}$ can be defined as an $N \times M$ matrix $\mathbf{G}^\dagger$, such that the solution estimate of $\mathbf{G}\mathbf{x} = \mathbf{d}$ can be expressed as $\mathbf{x} = \mathbf{G}^\dagger \mathbf{d}$. The necessary and sufficient condition for $\mathbf{G}^\dagger$ exists is

$$\text{If, and only if, } \quad \mathbf{G}\mathbf{G}^\dagger \mathbf{G} = \mathbf{G}. \tag{4.1}$$

SVD may play a fundamental role in the calculation and analysis of such a pseudo matrix inverse, $\mathbf{G}^\dagger$.

Eigenvalue analysis concerns the properties of a square matrix $\mathbf{A}$. Singular value analysis is a natural generalisation of the eigenvalue analysis of an arbitrary rectangular matrix $\mathbf{G}$.

4.1 Eigenvalues and eigenvectors

For a square matrix $\mathbf{A}$, there exists a special class of vectors, called eigenvectors. When any of these vectors is multiplied by matrix $\mathbf{A}$, it does not change the direction, but simply changes the scale:

$$\mathbf{A}\mathbf{x} = \lambda \mathbf{x}, \tag{4.2}$$

where the scalar λ is called the eigenvalue, associated with the eigenvector $\mathbf{x}$ of the matrix $\mathbf{A}$.

Seismic Inversion: Theory and Applications, First Edition. Yanghua Wang.
© 2017 John Wiley & Sons, Ltd. Published 2017 by John Wiley & Sons, Ltd.

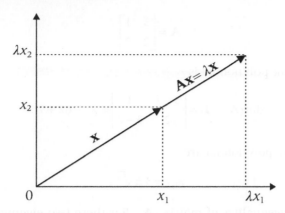

Figure 4.1 If **x** is an eigenvector of matrix **A**, matrix **A** acts to stretch the vector **x** and does not change its direction. The scale λ is the associated eigenvalue.

For solving the eigenvalues and associated eigenvectors, Equation 4.2 may be rearranged as

$$[\mathbf{A} - \lambda \mathbf{I}]\mathbf{x} = \mathbf{0} , \qquad (4.3)$$

where **I** is the identity matrix, and **0** is a null vector. Since **x** is nonzero, it means that matrix $[\mathbf{A} - \lambda \mathbf{I}]$ is singular,

$$\det(\mathbf{A} - \lambda \mathbf{I}) = 0 . \qquad (4.4)$$

This determinant is a polynomial in λ, called the characteristic polynomial. The roots of the characteristic polynomial are the eigenvalues of matrix **A**.

The eigenvalues are often sorted in decreasing absolute value as

$$|\lambda_1| \ge |\lambda_2| \ge \cdots \ge |\lambda_N| . \qquad (4.5)$$

If r is the rank of the $N \times N$ matrix **A**, the nonzero eigenvalues lie in the positions 1 to r, and the zero eigenvalues lie in the positions $r+1$ to N.

For each eigenvalue, solving Equation 4.3 may produce a set of eigenvectors, and the number of eigenvectors in the set associated with a single eigenvalue is infinite. This is because Equation 4.3 is a homogeneous system: if in system $f(\mathbf{x}) = 0$, then $f(\mathbf{x}) = f(c\mathbf{x}) = 0$, where c is a constant.

Let us demonstrate the calculation of eigenvalues and eigenvectors by an example. For a 2×2 matrix

$$\mathbf{A} = \begin{bmatrix} 3 & 1 \\ 2 & 3 \end{bmatrix},$$

the characteristic polynomial is

$$\det(\mathbf{A} - \lambda\mathbf{I}) = \begin{vmatrix} 3-\lambda & 1 \\ 2 & 3-\lambda \end{vmatrix} = \lambda^2 - 6\lambda + 7 .$$

The roots of this polynomial are

$$\lambda_{1,2} = 3 \pm \sqrt{2} .$$

They are the eigenvalues of matrix $\mathbf{A}$. For these two eigenvalues, solving the following two homogeneous systems may produce the associated eigenvectors:

$$\begin{bmatrix} -\sqrt{2} & 1 \\ 2 & -\sqrt{2} \end{bmatrix}\begin{bmatrix} x_1 \\ x_2 \end{bmatrix} = 0, \quad \text{and} \quad \begin{bmatrix} \sqrt{2} & 1 \\ 2 & \sqrt{2} \end{bmatrix}\begin{bmatrix} x_1 \\ x_2 \end{bmatrix} = 0 .$$

For any value of x_1, there is a value for x_2. If we set $x_1 = 1$, then two eigenvectors are derived:

$$\begin{bmatrix} 1 \\ \sqrt{2} \end{bmatrix}, \quad \text{and} \quad \begin{bmatrix} 1 \\ -\sqrt{2} \end{bmatrix} .$$

Then, normalisations lead to

$$\frac{1}{\sqrt{3}}\begin{bmatrix} 1 \\ \sqrt{2} \end{bmatrix}, \quad \text{and} \quad \frac{1}{\sqrt{3}}\begin{bmatrix} 1 \\ -\sqrt{2} \end{bmatrix} .$$

These normalised eigenvectors are unique.

Eigenvectors corresponding to different eigenvalues are linearly independent from one another. For the example here, only when $(a_1, a_2) = (0, 0)$, the following combination exists

$$a_1\begin{bmatrix} 1 \\ \sqrt{2} \end{bmatrix} + a_2\begin{bmatrix} 1 \\ -\sqrt{2} \end{bmatrix} = 0 .$$

Although the definition and also the exercise of the eigenvalue problem are straightforward, it is very difficult to algebraically compute the roots of any high order characteristic polynomial. In fact, the eigenvalue problem is much harder than solving $\mathbf{Ax} = \mathbf{b}$. An essential computational tool is the matrix diagonalisation.

4.2 Singular value concept

The eigenvalue concept for a square matrix can be extended to be singular values, worked on a rectangular matrix $\mathbf{G}$. The singular values of $\mathbf{G}$ are defined as the square roots of nonzero eigenvalues of $\mathbf{G}^T\mathbf{G}$ or $\mathbf{G}\mathbf{G}^T$.

As $\mathbf{G}$ is an $M \times N$ rectangular matrix, we can build a square matrix as

$$\begin{bmatrix} \mathbf{0} & \mathbf{G} \\ \mathbf{G}^T & \mathbf{0} \end{bmatrix}, \tag{4.6}$$

and conduct the eigenvalue analysis on both $\mathbf{G}$ and $\mathbf{G}^T$ simultaneously. For this expanded $(M+N)\times(N+M)$ square matrix, the corresponding eigenvalue problem becomes

$$\begin{bmatrix} \mathbf{0} & \mathbf{G} \\ \mathbf{G}^T & \mathbf{0} \end{bmatrix} \begin{bmatrix} \mathbf{u} \\ \mathbf{v} \end{bmatrix} = \lambda \begin{bmatrix} \mathbf{u} \\ \mathbf{v} \end{bmatrix}, \tag{4.7}$$

where $\mathbf{u}$ is an $M \times 1$ vector, and $\mathbf{v}$ is an $N \times 1$ vector.

Separating Equation 4.7 into two systems of equations yields

$$\mathbf{G}\mathbf{v} = \lambda\mathbf{u},$$
$$\mathbf{G}^T\mathbf{u} = \lambda\mathbf{v}. \tag{4.8}$$

It clearly indicates that if $(\lambda, \mathbf{u}, \mathbf{v})$ is a solution of Equation 4.7, so are $(-\lambda, -\mathbf{u}, \mathbf{v})$ and $(-\lambda, \mathbf{u}, -\mathbf{v})$ also a solution of Equation 4.7. For Equation 4.8, multiplying the first system of equations with $\mathbf{G}^T$ and the second system of equations with $\mathbf{G}$, we obtain

$$\mathbf{G}^T\mathbf{G}\mathbf{v} = \lambda^2\mathbf{v},$$
$$\mathbf{G}\mathbf{G}^T\mathbf{u} = \lambda^2\mathbf{u}. \tag{4.9}$$

It reveals that the $N \times N$ square matrix $\mathbf{G}^T\mathbf{G}$ and the $M \times M$ square matrix $\mathbf{G}\mathbf{G}^T$ have the same nonzero eigenvalues, λ^2.

Square-roots of these nonzero eigenvalues are called the singular values of the rectangular matrix $\mathbf{G}$. If $\text{rank}(\mathbf{G}) = r$, there are exactly r nonzero singular values:

$$\lambda_1 \geq \lambda_2 \geq \cdots \geq \lambda_r > 0. \tag{4.10}$$

For the $N \times N$ square matrix $\mathbf{G}^T\mathbf{G}$, there exist r vectors $[\mathbf{v}_1, \mathbf{v}_2, \cdots, \mathbf{v}_r]$ corresponding to the r non-vanishing eigenvalues, and $N-r$ vectors corresponding to the zero eigenvalues. For the $M \times M$ symmetric matrix $\mathbf{GG}^T$, there exist r vectors $[\mathbf{u}_1, \mathbf{u}_2, \cdots, \mathbf{u}_r]$ corresponding to the r non-vanishing eigenvalues, and $M - r$ vectors corresponding to the zero eigenvalues.

Let $\mathbf{U} = [\mathbf{u}_1, \mathbf{u}_2, \cdots, \mathbf{u}_M]$ be M eigenvectors of $\mathbf{GG}^T$, $\mathbf{V} = [\mathbf{v}_1, \mathbf{v}_2, \cdots, \mathbf{v}_N]$ be N eigenvectors of $\mathbf{G}^T\mathbf{G}$, and $\mathbf{\Lambda}$ be an $M \times N$ rectangular matrix with the singular values on its main diagonal and zero elsewhere, we can rewrite Equation 4.8 in a matrix form as

$$\mathbf{GV} = \mathbf{U}\mathbf{\Lambda},$$
$$\mathbf{G}^T\mathbf{U} = \mathbf{V}\mathbf{\Lambda}^T, \tag{4.11}$$

which gives the Lanczos representation (Lanczos, 1950) of an arbitrary rectangular matrix:

$$\mathbf{G} = \mathbf{U}\mathbf{\Lambda}\mathbf{V}^T. \tag{4.12}$$

This is a direct approach to the singular value decomposition (SVD) of any real $M \times N$ matrix $\mathbf{G}$:

1) In the $M \times N$ matrix $\mathbf{\Lambda}$, non-vanishing elements are only the nonzero singular values ranking from the position 1 to the position r on the principal diagonal.

2) Both matrices $\mathbf{U}$ and $\mathbf{V}$ are orthonormal whose columns (or rows) are orthogonal unit vectors:

$$\mathbf{U}^T\mathbf{U} = \mathbf{I}_{M \times M}, \qquad \mathbf{V}^T\mathbf{V} = \mathbf{I}_{N \times N}. \tag{4.13}$$

They can be verified by simply substituting $\mathbf{G}$ in Equation 4.11 with SVD expression of Equation 4.12.

The characteristic equation method of determining eigenvalues given in section 4.1 is not suitable for SVD implementation. Most practical algorithms for SVD are transformation-based methods, among which one of stable techniques is a two-phase method: (1) transforming the matrix to a bidiagonal form, and (2) performing QR decomposition to find the singular values of the bidiagonal matrix (Golub and Kahan, 1965; Golub and Reinsch, 1970). This two-phase, transformation-based algorithm is summarised in Appendix B.

4.3 Generalised inverse solution by SVD

Using SVD, $\mathbf{G} = \mathbf{U}\boldsymbol{\Lambda}\mathbf{V}^{\mathrm{T}}$, a pseudo-inverse (Moore, 1920; Penrose, 1955) of the rectangular matrix $\mathbf{G}$ can be given by

$$\mathbf{G}^{\dagger} = \mathbf{V}\boldsymbol{\Lambda}^{-1}\mathbf{U}^{\mathrm{T}}, \tag{4.14}$$

where

$$\boldsymbol{\Lambda}^{-1} = \begin{bmatrix} \lambda_1^{-1} & & & & & \\ & \lambda_2^{-1} & & & & \\ & & \ddots & & & \\ & & & \lambda_r^{-1} & & \\ & & & & 0 & \\ & & & & & \ddots \\ & & & & & & 0 \end{bmatrix}. \tag{4.15}$$

For a rectangular matrix $\mathbf{G}$, if $\mathrm{rank}(\mathbf{G}) = r$, and $\lambda_1 \geq \lambda_2 \geq \cdots \geq \lambda_r > 0$, set $\lambda_k^{-1} = 0$, for $\lambda_k = 0$, $k > r$. Therefore, $\mathrm{rank}(\mathbf{G}^{\dagger}) = \mathrm{rank}(\boldsymbol{\Lambda}^{-1}) = \mathrm{rank}(\boldsymbol{\Lambda}) = \mathrm{rank}(\mathbf{G}) = r$.

This pseudo-inverse $\mathbf{G}^{\dagger}$ is a generalised inverse. If $\mathbf{G}$ is full rank, the generalised inverse solution can be one of the following:

$$\mathbf{G}^{\dagger} = [\mathbf{G}^{\mathrm{T}}\mathbf{G}]^{-1}\mathbf{G}^{\mathrm{T}},$$
$$\mathbf{G}^{\dagger} = \mathbf{G}^{\mathrm{T}}[\mathbf{G}\mathbf{G}^{\mathrm{T}}]^{-1}. \tag{4.16}$$

The first equation is the least-squares solution and the second equation, which will be discussed below, is a least-norm solution. Substituting $\mathbf{G} = \mathbf{U}\boldsymbol{\Lambda}\mathbf{V}^{\mathrm{T}}$ into these two solutions, and considering $\mathbf{U}^{\mathrm{T}}\mathbf{U} = \mathbf{I}_{M \times M}$, $\mathbf{V}\mathbf{V}^{\mathrm{T}} = \mathbf{I}_{N \times N}$, we can obtain a unified solution $\mathbf{G}^{\dagger} = \mathbf{V}\boldsymbol{\Lambda}^{-1}\mathbf{U}^{\mathrm{T}}$. Hence, we refer to the pseudo-inverse $\mathbf{G}^{\dagger}$ as a generalised inverse solution.

The least-norm problem is to minimise $\sum_{j=0}^{N} x_j^2$, subject to the constraint that $\mathbf{e} = \mathbf{d} - \mathbf{G}\mathbf{x} = \mathbf{0}$. The objective function is

$$\phi(\mathbf{x}, \lambda) = \sum_{j=0}^{N} x_j^2 + \sum_{i=1}^{M} \lambda_i e_i = \mathbf{x}^{\mathrm{T}}\mathbf{x} + \lambda^{\mathrm{T}}(\mathbf{d} - \mathbf{G}\mathbf{x}), \tag{4.17}$$

where λ_i are the Lagrange multipliers. Setting $\partial\phi / \partial\mathbf{x} = 0$ yields

$$\mathbf{x} = \frac{1}{2}\mathbf{G}^{\mathrm{T}}\lambda. \tag{4.18}$$

Setting $\partial\phi/\partial\lambda = 0$ yields $\mathbf{d} - \mathbf{Gx} = \mathbf{0}$, which is the constraint equation. Inserting expression $\mathbf{x} = \frac{1}{2}\mathbf{G}^T\boldsymbol{\lambda}$ into the constraint equation, $\mathbf{d} - \mathbf{Gx} = \mathbf{0}$, leads to

$$\mathbf{GG}^T\boldsymbol{\lambda} = 2\mathbf{d}\ . \tag{4.19}$$

Therefore, the solution vector $\boldsymbol{\lambda}$ is

$$\boldsymbol{\lambda} = 2[\mathbf{GG}^T + \mu\mathbf{I}]^{-1}\mathbf{d}\ , \tag{4.20}$$

and then the solution $\mathbf{x}$ is

$$\mathbf{x} = \mathbf{G}^T[\mathbf{GG}^T + \mu\mathbf{I}]^{-1}\mathbf{d}\ . \tag{4.21}$$

When $\mu = 0$, this is the second generalised inverse $\mathbf{G}^\dagger$ in Equation 4.16.

Therefore, if $\mathbf{G}$ is not full rank, stabilised solutions of either least-squares or least-norm problems are

$$\mathbf{G}^\dagger = [\mathbf{G}^T\mathbf{G} + \mu\mathbf{I}]^{-1}\mathbf{G}^T,$$
$$\mathbf{G}^\dagger = \mathbf{G}^T[\mathbf{GG}^T + \mu\mathbf{I}]^{-1}\ . \tag{4.22}$$

The first least-squares solution is applicable to the so-called tall-and-skinny matrix $\mathbf{G}$, and the second least-norm solution is applicable to the so-called fat matrix $\mathbf{G}$, for an underdetermined problem. Substituting $\mathbf{G} = \mathbf{U}\boldsymbol{\Lambda}\mathbf{V}^T$ into the two solutions in Equation 4.22 will produce a unified solution as in the following:

$$\mathbf{G}^\dagger = \mathbf{V}\frac{\boldsymbol{\Lambda}}{\boldsymbol{\Lambda}^2 + \mu\mathbf{I}}\mathbf{U}^T\ . \tag{4.23}$$

4.4 SVD applications

With the generalised inverse $\mathbf{G}^\dagger$, we define a resolution matrix for the solution or model estimate as

$$\mathbf{R} \equiv \mathbf{G}^\dagger\mathbf{G} = \mathbf{V}\mathbf{V}^T. \tag{4.24}$$

To understand this concept, let us see the solution

$$\tilde{\mathbf{x}} = \mathbf{G}^\dagger\tilde{\mathbf{d}}\ , \tag{4.25}$$

and perform a manipulation, as in the following

$$\tilde{\mathbf{x}} = \mathbf{G}^\dagger\tilde{\mathbf{d}} + (\mathbf{G}^\dagger\mathbf{d} - \mathbf{G}^\dagger\mathbf{d}) = \mathbf{G}^\dagger\mathbf{d} + \mathbf{G}^\dagger(\tilde{\mathbf{d}} - \mathbf{d}) = \mathbf{R}\mathbf{x} + \mathbf{G}^\dagger(\tilde{\mathbf{d}} - \mathbf{d})\ .$$
$$\tag{4.26}$$

The discrepancy between the solution estimate $\tilde{\mathbf{x}}$ and the true model $\mathbf{x}$ is

$$\tilde{\mathbf{x}} - \mathbf{x} = (\mathbf{R} - \mathbf{I})\mathbf{x} + \mathbf{G}^{\dagger}(\tilde{\mathbf{d}} - \mathbf{d}) . \tag{4.27}$$

Therefore, the resolution of the solution estimate for the ith model parameter is defined by the closeness between the ith row of $\mathbf{R}$ and the δ function. In other words, the deviation of $\mathbf{R}$ from the identity matrix $\mathbf{I}$ determines the degree of distrust we should have in the components of the solution vectors $\tilde{\mathbf{x}}$ that are most poorly resolved.

In the absence of data errors, $\tilde{\mathbf{x}} = \mathbf{R}\mathbf{x}$. Hence, $\mathbf{R}$ represents a linear mapping between the true solution and the model estimate. Only when $\mathbf{R} = \mathbf{I}$, the solution discrepancy $\tilde{\mathbf{x}} - \mathbf{x}$ is caused purely by the data errors $\tilde{\mathbf{d}} - \mathbf{d}$.

We can define a similar resolution matrix for the data as

$$\mathbf{P} \equiv \mathbf{G}\mathbf{G}^{\dagger} = \mathbf{U}\mathbf{U}^{\mathrm{T}} . \tag{4.28}$$

It is called also the information density matrix. To understand this concept, let us see the solution $\tilde{\mathbf{x}} = \mathbf{G}^{\dagger}\tilde{\mathbf{d}}$ again. Pre-multiplying $\mathbf{G}\mathbf{G}^{\dagger}$ to both sides,

$$\mathbf{G}\mathbf{G}^{\dagger}\mathbf{G}\tilde{\mathbf{x}} = \mathbf{G}\mathbf{G}^{\dagger}\tilde{\mathbf{d}} , \tag{4.29}$$

and denoting $\mathbf{G}\mathbf{G}^{\dagger}\mathbf{G} = \mathbf{G}$ in the left-hand side and $\mathbf{P} = \mathbf{G}\mathbf{G}^{\dagger}$ in the right hand side, we have the following expression

$$\mathbf{G}\tilde{\mathbf{x}} = \mathbf{P}\tilde{\mathbf{d}} . \tag{4.30}$$

In this equation, where the synthetic data are represented by the left-hand side, the right-hand side is a filtered version of the real observed data. The filter is $\mathbf{P}$. In other words, $\mathbf{P}$ measures the fit between real data and synthetic data.

The covariance concept is a measure of how much two variables change together. If two variables tend to vary together (that is, when one of them is above its expected value, then the other variable tends to be above its expected value too), then the covariance between the two variables will be positive. On the other hand, if one of them tends to be above its expected value when the other variable is below its expected value, then the covariance between the two variables will be negative. Let us see the model covariance matrix $\mathbf{C}_{\mathrm{x}}$:

$$\mathbf{C}_{\mathrm{x}} = \tilde{\mathbf{x}}\,\tilde{\mathbf{x}}^{\mathrm{T}} . \tag{4.31}$$

Substituting $\tilde{\mathbf{x}} = \mathbf{V}\boldsymbol{\Lambda}^{-1}\mathbf{U}^{\mathrm{T}}\tilde{\mathbf{d}}$, we have

$$\mathbf{C}_{\mathrm{x}} = \mathbf{V}\boldsymbol{\Lambda}^{-1}\mathbf{U}^{\mathrm{T}}\mathbf{C}_{\mathrm{d}}\mathbf{U}\boldsymbol{\Lambda}^{-1}\mathbf{V}^{\mathrm{T}}, \tag{4.32}$$

where $\mathbf{C}_{\mathrm{d}} = \tilde{\mathbf{d}}\,\tilde{\mathbf{d}}^{\mathrm{T}}$ is the data covariance matrix. Assuming $\mathbf{C}_{\mathrm{d}} = \sigma^2\mathbf{I}$, where σ^2 is the data variance, we obtain

$$\mathbf{C}_{\mathrm{x}} = \sigma^2\boldsymbol{\Lambda}^{-2}. \tag{4.33}$$

Before we discuss further on the applications of SVD, we need to understand the concept of *matrix norm*. In seismic inversion quite often we shall be concerned with norms of vectors and norms of matrices at the same time. Thus, it will be convenient to have a matrix norm induced by the vector norm.

Recall that the norm of a vector $\mathbf{x} = [x_1, x_2, \cdots, x_N]^{\mathrm{T}}$ is defined by

$$L_p \equiv \parallel \mathbf{x} \parallel_p = (\mid x_1 \mid^p + \mid x_2 \mid^p + \cdots \mid x_N \mid^p)^{1/p}. \tag{4.34}$$

Hence, the L_1 norm is the sum of the absolute values of the vector components, and the L_2 norm is the square root of the sum of squares of its components. Suppose $X \equiv \max \mid x_k \mid$ and there are K components which have the same absolute value X. Then, following Equation 4.34,

$$\lim_{p\to\infty} \parallel \mathbf{x} \parallel_p = \lim_{p\to\infty}\left(\frac{\mid x_1 \mid^p}{X^p} + \frac{\mid x_2 \mid^p}{X^p} + \cdots \frac{\mid x_N \mid^p}{X^p} \right)^{1/p} X = \lim_{p\to\infty} K^{1/p} X = X, \tag{4.35}$$

we have the L_∞ norm (the infinity norm) definition as $\parallel \mathbf{x} \parallel_\infty = \max \mid x_k \mid$.

The definition of the induced norm of a matrix, associated with the vector norm, is given by

$$\parallel \mathbf{G} \parallel_p \equiv \max_{\mathbf{x}\neq 0}\left(\frac{\parallel \mathbf{G}\mathbf{x} \parallel_p}{\parallel \mathbf{x} \parallel_p} \right). \tag{4.36}$$

It follows immediately from this definition that

$$\parallel \mathbf{G}\mathbf{x} \parallel_p \leq \parallel \mathbf{G} \parallel_p \parallel \mathbf{x} \parallel_p. \tag{4.37}$$

If the elements of matrix $\mathbf{G}$ are g_{ij}, the L_1 norm is the maximum value among the sums of the absolute values of the columns:

$$\parallel \mathbf{G} \parallel_1 = \max_{1\leq j\leq N}\sum_{i=1}^{N} \mid g_{ij} \mid, \tag{4.38}$$

and the L_∞ norm is the maximum value among the sums of the absolute values of the rows:

$$\| \mathbf{G} \|_\infty = \max_{1 \le i \le N} \sum_{j=1}^{N} | g_{ij} | . \tag{4.39}$$

For the L_2 norm, in theory we have

$$\| \mathbf{G} \|_2 = \max_{\|\mathbf{x}\|_2 = 1} \| \mathbf{Gx} \|_2 . \tag{4.40}$$

Note that, in fact, this L_2 norm is not very easy to calculate since one must examine all possible vectors of length 1 and find the maximum possible result. Alternatively, it may be evaluated based on the singular value of matrix $\mathbf{G}$: The matrix norm $\| \mathbf{G} \|_2$ is the largest eigenvalue of matrix $\mathbf{G}$.

The three matrix norms $\| \mathbf{G} \|_1$, $\| \mathbf{G} \|_2$ and $\| \mathbf{G} \|_\infty$ satisfy the inequality

$$\| \mathbf{G} \|_2^2 \le \| \mathbf{G} \|_1 \| \mathbf{G} \|_\infty . \tag{4.41}$$

The matrix norm $\| \mathbf{G} \|_2$ is often used, and is popularly denoted as $\| \mathbf{G} \|$.

The following properties of the SVD are also important in applications.

1) Any orthogonal transform does not change the singular values of a matrix.

2) The L_2 norm of matrix $\mathbf{G}$ equals the L_2 norm of matrix $\boldsymbol{\Lambda}$, that is, $\| \mathbf{G} \| = \| \boldsymbol{\Lambda} \|$.

3) The largest and the smallest singular values are

$$\lambda_1 = \max_{\mathbf{x} \ne 0} \frac{\| \mathbf{Gx} \|}{\| \mathbf{x} \|} , \qquad \lambda_N = \min_{\mathbf{x} \ne 0} \frac{\| \mathbf{Gx} \|}{\| \mathbf{x} \|} . \tag{4.42}$$

Therefore, it is easy to see that the following two L_2 norms of matrices are related to the singular values as

$$\| \mathbf{G} \| = \lambda_1 , \qquad \| \mathbf{G}^{-1} \| = \frac{1}{\lambda_N} . \tag{4.43}$$

4) The condition number of a matrix equals the ratio of the largest eigenvalue to the smallest eigenvalue:

$$\text{cond}(\mathbf{G}) = \frac{\lambda_1}{\lambda_N} . \tag{4.44}$$

Let vector $\mathbf{e}$ be the error in data vector $\mathbf{d}$, then the error in the solution $(\mathbf{x} = \mathbf{G}^{-1}\mathbf{d})$ is $\mathbf{G}^{-1}\mathbf{e}$. The condition number is defined in terms of the relative error in $\mathbf{m}$ divided by the relative error in $\mathbf{d}$, as in the following

$$\text{cond}(\mathbf{G}) = \max\left(\frac{\|\mathbf{G}^{-1}\mathbf{e}\| / \|\mathbf{G}^{-1}\mathbf{d}\|}{\|\mathbf{e}\| / \|\mathbf{d}\|}\right)$$

$$= \max\left(\frac{\|\mathbf{d}\|}{\|\mathbf{G}^{-1}\mathbf{d}\|}\right)\max\left(\frac{\|\mathbf{G}^{-1}\mathbf{e}\|}{\|\mathbf{e}\|}\right) \tag{4.45}$$

$$= \|\mathbf{G}\| \cdot \|\mathbf{G}^{-1}\|.$$

As $\|\mathbf{G}\| = \lambda_1$ and $\|\mathbf{G}^{-1}\| = \lambda_N^{-1}$, we obtain the expression in Equation 4.44. As it is defined by the product of norms of two operators $\mathbf{G}$ and its inverse $\mathbf{G}^{-1}$, the condition number reflects the property of the matrix $\mathbf{G}$. However, we should think of the condition number as being (very roughly) the rate at which the solution $\mathbf{x}$ will change with respect to a change in $\mathbf{d}$. The condition number gives a measure in order to see if the problem is ill-conditioned, which corresponds to a large condition number (Franklin, 1970). In such cases, the generalised solution in Equation 4.25 can improve the solution estimate by damping the small singular values λ_k.

Damping (or simply zeroing) a singular value means to reduce the effect of (or eliminating) one linear combination of the set of equations that we are trying to solve. Thus, the procedure does usually reduce the resolution of the solution estimate.

CHAPTER 5

Gradient-based methods

If the system of linear equations is large, instead of using a direct solution as presented in the previous chapter, it can be solved iteratively. An iterative method starts with an approximation $\mathbf{x}^{(k)}$ and refines successively the solution estimate in the form $\mathbf{x}^{(k+1)} = f(\mathbf{x}^{(k)})$, where f is an iteration function, and k in parentheses is the iteration number.

For understanding the iterative procedure, let us try one such implementation on the linear system $\mathbf{A}\mathbf{x} = \mathbf{b}$, where $\mathbf{A}$ is a non-singular $N \times N$ square matrix. We can introduce another non-singular square matrix $\mathbf{B}$ arbitrarily so that

$$\mathbf{B}\mathbf{x} + (\mathbf{A} - \mathbf{B})\mathbf{x} = \mathbf{b}. \tag{5.1}$$

Then, by making the *ansatz*, an educated approximation that will get verified via its results,

$$\mathbf{B}\mathbf{x}^{(k+1)} + (\mathbf{A} - \mathbf{B})\mathbf{x}^{(k)} = \mathbf{b}, \tag{5.2}$$

we have

$$\mathbf{x}^{(k+1)} = \mathbf{x}^{(k)} + \mathbf{B}^{-1}\mathbf{e}^{(k)}, \tag{5.3}$$

where $\mathbf{e}^{(k)} = \mathbf{b} - \mathbf{A}\mathbf{x}^{(k)}$ is the vector of residuals. The calculation of inverse matrix $\mathbf{B}^{-1}$ can still be time consuming. Thus, we can approximate it further to be a constant, and express the solution estimate $\mathbf{x}^{(k+1)}$ simply as

$$\mathbf{x}^{(k+1)} = \mathbf{x}^{(k)} + \alpha_k \mathbf{e}^{(k)}. \tag{5.4}$$

Considering vector $\mathbf{e}^{(k)}$ as an updating direction, we may express this updating equation in a general form as

Seismic Inversion: Theory and Applications, First Edition. Yanghua Wang.
© 2017 John Wiley & Sons, Ltd. Published 2017 by John Wiley & Sons, Ltd.

$$\mathbf{x}^{(k+1)} = \mathbf{x}^{(k)} + \alpha_k \mathbf{v}^{(k)} , \qquad (5.5)$$

where $\mathbf{v}^{(k)}$ is the updating direction, and α_k is the step length.

For each individual iteration, we shall treat it generally as an inverse problem: defining an objective function, and updating the solution estimate based on the gradient of the objective function. Because the updating direction $\mathbf{v}^{(k)}$ is defined in terms of the gradient vector, $\boldsymbol{\gamma}^{(k)}$, it is called the gradient-based method.

We shall see that the vector $\mathbf{e}^{(k)}$ in Equation 5.4 is related to the negative gradient vector $\boldsymbol{\gamma}^{(k)}$ of an error function. Then, we have a sample gradient-based method, as the following

$$\mathbf{x}^{(k+1)} = \mathbf{x}^{(k)} - \alpha_k \boldsymbol{\gamma}^{(k)} . \qquad (5.6)$$

Here the updating direction $\mathbf{v}^{(k)}$ is the negative gradient vector $\boldsymbol{\gamma}^{(k)}$. It is called the steepest decent method, that will be discussed in section 5.2.

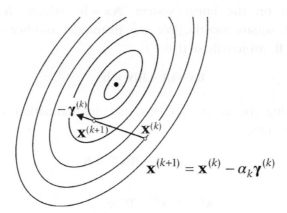

Figure 5.1 Given a gradient vector $\boldsymbol{\gamma}^{(k)}$ at $\mathbf{x}^{(k)}$, search for the local minimum position $\mathbf{x}^{(k+1)}$ along the negative gradient direction, $-\boldsymbol{\gamma}^{(k)}$. The step length α_k is originated from the trial solution $\mathbf{x}^{(k)}$.

5.1 The step length

For each individual iteration, we define an objective function, using an error function, $\phi(\mathbf{e})$, where $\mathbf{e}$ is a residual vector for a given trial solution $\mathbf{x}$, and then solve this inverse problem based on a minimal variation principle.

Given the residual vector $\mathbf{e} = \mathbf{b} - \mathbf{A}\mathbf{x}$, its projection onto the variable vector $\mathbf{x}$ is $\mathbf{A}^{-1}\mathbf{e}$. The error function can be defined by the inner product

$$\phi(\mathbf{e}) = (\mathbf{e}, \mathbf{A}^{-1}\mathbf{e}). \tag{5.7}$$

In seismic inversion, $\mathbf{e}$ is the vector of data residuals in the data space, and $\mathbf{A}^{-1}\mathbf{e} = \mathbf{A}^{-1}\mathbf{b} - \mathbf{x}$ is the vector of model errors in the model space. Minimising the error function $\phi(\mathbf{e})$ means to minimise the data residuals and the model errors, simultaneously.

Assuming that matrix $\mathbf{A}$ is symmetric, $\mathbf{A}^{T} = \mathbf{A}$, then substituting $\mathbf{e} = \mathbf{b} - \mathbf{A}\mathbf{x}$ into the error function of Equation 5.7 leads to

$$\phi(\mathbf{x}) = (\mathbf{x}, \mathbf{A}\mathbf{x}) - 2(\mathbf{b}, \mathbf{x}) + (\mathbf{b}, \mathbf{A}^{-1}\mathbf{b}). \tag{5.8}$$

Because this error function $\phi(\mathbf{x})$ is quadratic in variable vector $\mathbf{x}$, if we set $\partial\phi(\mathbf{x})/\partial\mathbf{x} = 0$, we have a linear system $\mathbf{A}\mathbf{x} - \mathbf{b} = 0$. But, we want to solve $\mathbf{x}$ iteratively, rather than directly.

At the kth iteration, solution estimate $\mathbf{x}^{(k)}$ may represent a point in N-dimensional space, and then the updating equation $\mathbf{x}^{(k+1)} = \mathbf{x}^{(k)} + \alpha_k \mathbf{v}^{(k)}$ defines a line between points $\mathbf{x}^{(k)}$ and $\mathbf{x}^{(k+1)}$. The error function $\phi(\mathbf{x}^{(k+1)})$ can be written as

$$\phi(\alpha_k) = \alpha_k^2(\mathbf{v}^{(k)}, \mathbf{A}\mathbf{v}^{(k)}) - 2\alpha_k(\mathbf{v}^{(k)}, \mathbf{e}^{(k)}) + (\mathbf{x}^{(k)}, \mathbf{A}\mathbf{x}^{(k)}) + (\mathbf{b}, \mathbf{A}^{-1}\mathbf{b}). \tag{5.9}$$

The derivations in Equations 5.8 and 5.9 have exploited the following properties of the inner product:

- symmetry $(\mathbf{r}_1, \mathbf{r}_2) = (\mathbf{r}_2, \mathbf{r}_1)$;
- linearity $(\mathbf{r}_1, \mathbf{r}_2 + \mathbf{r}_3) = (\mathbf{r}_1, \mathbf{r}_2) + (\mathbf{r}_1, \mathbf{r}_3)$; and
- continuity $c(\mathbf{r}_1, \mathbf{r}_2) = (c\mathbf{r}_1, \mathbf{r}_2) = (\mathbf{r}_1, c\mathbf{r}_2)$, where c is a constant.

In addition, matrix $\mathbf{A}$ is assumed to be symmetric, $\mathbf{A}^{T} = \mathbf{A}$. For a general rectangular matrix $\mathbf{G}$ in seismic inversion, pre-multiplying its transpose can produce such a symmetric matrix $\mathbf{A} = \mathbf{G}^{T}\mathbf{G}$.

Equation 5.9 indicates that the error function $\phi(\alpha_k)$ is a quadratic function of α_k. A local minima along the line $\mathbf{x}^{(k)} \rightarrow \mathbf{x}^{(k+1)}$ can be found by setting

$$\frac{\partial\phi}{\partial\alpha_k} = 2\alpha_k(\mathbf{v}^{(k)}, \mathbf{A}\mathbf{v}^{(k)}) - 2(\mathbf{v}^{(k)}, \mathbf{e}^{(k)}) = 0. \tag{5.10}$$

Therefore, the step length for the updating the solution estimate is

$$\alpha_k = \frac{(\mathbf{v}^{(k)}, \mathbf{e}^{(k)})}{(\mathbf{v}^{(k)}, \mathbf{A}\mathbf{v}^{(k)})}. \qquad (5.11)$$

Different gradient-based methods, such as the steepest descent method here and the conjugate gradient method that follows, only differ in the choice of directions $\mathbf{v}^{(k)}$.

5.2 The steepest descent method

The steepest descent method chooses the residual vector as the updating direction $\mathbf{v}^{(k)} = \mathbf{e}^{(k)}$, as shown in Equation 5.4. For the steepest descent method, we need to understand the following four aspects.

(1) It is indeed a gradient-based method. The idea behind the steepest descent method is to minimise $\phi(\mathbf{x}^{(k)})$ repeatedly along lines defined by the residual vector $\mathbf{e}^{(k)}$. According to the error function of Equation 5.8, after k iterations, the error function is

$$\phi(\mathbf{x}^{(k)}) = (\mathbf{x}^{(k)}, \mathbf{A}\mathbf{x}^{(k)}) - 2(\mathbf{b}, \mathbf{x}^{(k)}) + (\mathbf{b}, \mathbf{A}^{-1}\mathbf{b}), \qquad (5.12)$$

and its gradient is

$$\phi'(\mathbf{x}^{(k)}) = 2(\mathbf{A}\mathbf{x}^{(k)} - \mathbf{b}). \qquad (5.13)$$

The residual vector $\mathbf{e}^{(k)}$ is proportional to the gradient,

$$\mathbf{e}^{(k)} = \mathbf{b} - \mathbf{A}\mathbf{x}^{(k)} = -\frac{1}{2}\phi'(\mathbf{x}^{(k)}). \qquad (5.14)$$

The '−' sign reveals that the steepest descent method updates the solution estimate along the negative gradient direction.

When $\phi'(\mathbf{x}^{(k)})$ vanishes, $\mathbf{e}^{(k)} = \mathbf{0}$, then $\mathbf{x}^{(k)}$ is the solution of equation $\mathbf{A}\mathbf{x} = \mathbf{b}$.

(2) If we take double the step length, $2\alpha_k$, we arrive at point $\mathbf{x}^{(k)} + 2\alpha_k\mathbf{e}^{(k)}$, which is on the same contour of the quadratic function $\phi(\mathbf{x}^{(k)})$:

$$\phi(\mathbf{x}^{(k)} + 2\alpha_k\mathbf{e}^{(k)}) = \phi(\mathbf{x}^{(k)}). \qquad (5.15)$$

Thus, point $\mathbf{x}^{(k+1)} = \mathbf{x}^{(k)} + \alpha_k\mathbf{e}^{(k)}$ is right in the middle between two points $\mathbf{x}^{(k)}$ and $\mathbf{x}^{(k)} + 2\alpha_k\mathbf{e}^{(k)}$ on the same contour $\phi(\mathbf{x}^{(k)})$.

If we want to slow down the convergence, for stabilisation, we can adjust the step length α_k by a relaxation factor. Equation 5.15 suggests that the relaxation factor should be within the range (0, 2). When the

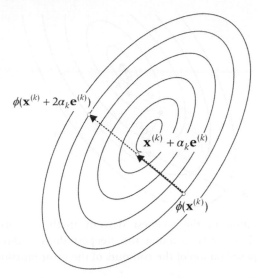

Figure 5.2 The steepest descent method: $\phi(\mathbf{x}^{(k)})$ and $\phi(\mathbf{x}^{(k)} + 2\alpha_k\mathbf{e}^{(k)})$ are on the same contour of the error function, and the steepest descent algorithm moves to $\mathbf{x}^{(k)} + \alpha_k\mathbf{e}^{(k)}$ the midpoint in between.

relaxation factor > 1, we might speed up the convergence or encounter an unstable issue.

(3) The residual vectors $\{\mathbf{e}^{(k)}, k = 1, 2, \cdots\}$ are pair-wise orthogonal. The updated residual vector $\mathbf{e}^{(k+1)}$ is

$$\mathbf{e}^{(k+1)} = \mathbf{b} - \mathbf{A}\mathbf{x}^{(k+1)} = \mathbf{b} - \mathbf{A}(\mathbf{x}^{(k)} + \alpha_k\mathbf{e}^{(k)})$$

$$= \mathbf{e}^{(k)} - \alpha_k\mathbf{A}\mathbf{e}^{(k)}. \tag{5.16}$$

The inner product $(\mathbf{e}^{(k)}, \mathbf{e}^{(k+1)})$ is

$$(\mathbf{e}^{(k)}, \mathbf{e}^{(k+1)}) = (\mathbf{e}^{(k)}, \mathbf{e}^{(k)}) - \alpha_k(\mathbf{e}^{(k)}, \mathbf{A}\mathbf{e}^{(k)})$$

$$= (\mathbf{e}^{(k)}, \mathbf{e}^{(k)}) - \frac{(\mathbf{e}^{(k)}, \mathbf{e}^{(k)})}{(\mathbf{e}^{(k)}, \mathbf{A}\mathbf{e}^{(k)})}(\mathbf{e}^{(k)}, \mathbf{A}\mathbf{e}^{(k)}) \tag{5.17}$$

$$= 0.$$

Therefore, the residual at each step of the steepest descent method is orthogonal to the residuals before and after. Figure 5.3 shows graphically the solution vector as a function of iteration, superposed onto contour plot of the error function $\phi(\mathbf{x})$. The steepest descent directions $\{\mathbf{e}^{(1)}, \mathbf{e}^{(2)}, \cdots\}$ are pair-wise orthogonal.

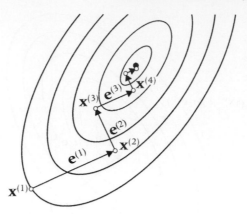

Figure 5.3   Illustration of the steepest descent method: Start from the trial solution $\mathbf{x}^{(1)}$, which is updated along the steepest descent direction $\mathbf{e}^{(1)}$, then $\mathbf{e}^{(2)}$, $\mathbf{e}^{(3)}$, $\cdots$, superposed on top of the contours of the error function $\phi(\mathbf{x})$.

(4) The steepest descent method is converged. The difference of the error functions between two consecutive iterations is

$$\phi(\mathbf{x}^{(k+1)}) - \phi(\mathbf{x}^{(k)}) = (\mathbf{x}^{(k+1)}, \mathbf{A}\mathbf{x}^{(k+1)}) - (\mathbf{x}^{(k)}, \mathbf{A}\mathbf{x}^{(k)}) - 2(\mathbf{b}, \mathbf{x}^{(k+1)} - \mathbf{x}^{(k)}).$$

$$(5.18)$$

Substituting $\mathbf{x}^{(k+1)} = \mathbf{x}^{(k)} + \alpha_k \mathbf{e}^{(k)}$, and assuming that $\mathbf{A}$ is symmetric, it becomes

$$\phi(\mathbf{x}^{(k+1)}) - \phi(\mathbf{x}^{(k)}) = \alpha_k^2(\mathbf{e}^{(k)}, \mathbf{A}\mathbf{e}^{(k)}) - 2\alpha_k(\mathbf{e}^{(k)}, \mathbf{e}^{(k)}). \qquad (5.19)$$

Then using definition of $\alpha_k = (\mathbf{e}^{(k)}, \mathbf{e}^{(k)})/(\mathbf{e}^{(k)}, \mathbf{A}\mathbf{e}^{(k)})$ leads to

$$\phi(\mathbf{x}^{(k+1)}) - \phi(\mathbf{x}^{(k)}) = -\frac{(\mathbf{e}^{(k)}, \mathbf{e}^{(k)})^2}{(\mathbf{e}^{(k)}, \mathbf{A}\mathbf{e}^{(k)})} \leq 0. \qquad (5.20)$$

Note here, the symmetric matrix $\mathbf{A}$ is assumed to be positive definite, that is, $(\mathbf{e}, \mathbf{A}\mathbf{e}) = (\mathbf{A}\mathbf{e}, \mathbf{e}) > 0$ for all nonzero real vectors $\mathbf{e}$. Since $\phi(\mathbf{x}^{(k+1)}) \leq \phi(\mathbf{x}^{(k)})$ for any $\mathbf{x}^{(k)}$, then

$$\phi(\mathbf{x}^{(1)}) \geq \phi(\mathbf{x}^{(2)}) \geq \cdots \geq \phi(\mathbf{x}^{(k)}) \geq \phi(\mathbf{x}^{(k+1)}) \geq \cdots \qquad (5.21)$$

is a monotone sequence, which must converge.

The implementation of the steepest descent method may be summarised as follows.

Choose $\mathbf{x}^{(1)}$, this gives $\mathbf{e}^{(1)} = \mathbf{b} - \mathbf{A}\mathbf{x}^{(1)}$;
Then, iteration for $k = 1, 2, 3, \cdots$:

$$\alpha_k = \frac{(\mathbf{e}^{(k)}, \mathbf{e}^{(k)})}{(\mathbf{e}^{(k)}, \mathbf{A}\mathbf{e}^{(k)})},$$

$$\mathbf{x}^{(k+1)} = \mathbf{x}^{(k)} + \alpha_k \mathbf{e}^{(k)}, \qquad (5.22)$$

$$\mathbf{e}^{(k+1)} = \mathbf{e}^{(k)} - \alpha_k \mathbf{A}\mathbf{e}^{(k)}.$$

For each iteration, this implementation involves only one matrix-vector production $\mathbf{A}\mathbf{e}^{(k)}$. Therefore, the steepest descent method is efficient if compared to any direct solution through the inverse matrix. The efficiency achieved at each step comes at the price of poor convergence rates.

5.3 Conjugate gradient method

The conjugate gradient method (Hestenes and Stiefel, 1952) defines the direction $\mathbf{v}$ by an auxiliary vector $\mathbf{p}$ which closes to the residual vector $\mathbf{e}$, but also holds the following conjugate property:

$$(\mathbf{p}^{(i)}, \mathbf{A}\mathbf{p}^{(j)}) = 0, \qquad \text{for } i \neq j. \qquad (5.23)$$

Note that this *conjugate* notion is not related to the complex conjugate. If two nonzero vectors $\mathbf{p}^{(i)}$ and $\mathbf{p}^{(j)}$ satisfy property of Equation 5.23, they are conjugate with respect to matrix $\mathbf{A}$, or $\mathbf{A}$-orthogonal. Assuming that $\mathbf{A}$ is symmetric, then if $\mathbf{p}^{(i)}$ is conjugate to $\mathbf{p}^{(j)}$, $\mathbf{p}^{(j)}$ is also conjugate to $\mathbf{p}^{(i)}$ with respect to matrix $\mathbf{A}$:

$$(\mathbf{p}^{(i)}, \mathbf{A}\mathbf{p}^{(j)}) = (\mathbf{A}^{\mathrm{T}}\mathbf{p}^{(i)}, \mathbf{p}^{(j)}) = (\mathbf{A}\mathbf{p}^{(i)}, \mathbf{p}^{(j)}). \qquad (5.24)$$

In contrast to the steepest descent method, in which the updating directions $\mathbf{v}^{(k)} \equiv \mathbf{e}^{(k)}$ are pair-wise orthogonal $(\mathbf{v}^{(k+1)}, \mathbf{v}^{(k)}) = 0$, the conjugate gradient method relaxes this condition of orthogonality and requires the updating directions $\mathbf{v}^{(k)} \equiv \mathbf{p}^{(k)}$ being conjugate or $\mathbf{A}$-orthogonal, $(\mathbf{v}^{(k+1)}, \mathbf{A}\mathbf{v}^{(k)}) = 0$. This relaxation will speed up the convergence.

The conjugate gradient method insists on taking a direction $\mathbf{p}^{(k+1)}$ closest to the vector $\mathbf{e}^{(k+1)}$, the negative gradient of the error function $\phi(\mathbf{x}^{(k+1)})$, under the conjugacy constraint, $(\mathbf{p}^{(k+1)}, \mathbf{A}\mathbf{p}^{(k)}) = 0$.

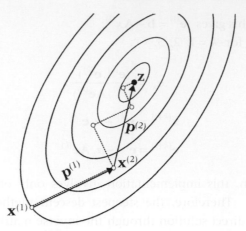

Figure 5.4 Solid lines illustrate solution updates in the conjugate gradient method, whereas dashed lines are the solution update vectors in the steepest descent method.

Defining the updating direction $\mathbf{p}^{(k+1)}$ as

$$\mathbf{p}^{(k+1)} = \mathbf{e}^{(k+1)} + \beta_k \mathbf{p}^{(k)}, \qquad (5.25)$$

where β_k is a coefficient chosen so as to guarantee the **A**-orthogonality of vectors $\mathbf{p}^{(k)}$ and $\mathbf{p}^{(k+1)}$, we have

$$(\mathbf{p}^{(k+1)}, \mathbf{Ap}^{(k)}) = (\mathbf{e}^{(k+1)}, \mathbf{Ap}^{(k)}) + \beta_k(\mathbf{p}^{(k)}, \mathbf{Ap}^{(k)}) = 0 . \qquad (5.26)$$

Then, we obtain the coefficient β_k as

$$\beta_k = -\frac{(\mathbf{e}^{(k+1)}, \mathbf{Ap}^{(k)})}{(\mathbf{p}^{(k)}, \mathbf{Ap}^{(k)})} . \qquad (5.27)$$

Using vector $\mathbf{p}^{(k)}$ to replace $\mathbf{v}^{(k)}$ in Equation 5.11, we have the step length α_k as

$$\alpha_k = \frac{(\mathbf{p}^{(k)}, \mathbf{e}^{(k)})}{(\mathbf{p}^{(k)}, \mathbf{Ap}^{(k)})}, \qquad (5.28)$$

for updating the solution estimate, $\mathbf{x}^{(k+1)} = \mathbf{x}^{(k)} + \alpha_k \mathbf{p}^{(k)}$.

As the **A**-orthogonal search vectors $\mathbf{p}^{(k)}$ are computed on the fly, the residuals can also be calculated recursively by

$$\mathbf{e}^{(k+1)} = \mathbf{b} - \mathbf{A}\mathbf{x}^{(k+1)} = \mathbf{b} - \mathbf{A}(\mathbf{x}^{(k)} + \alpha_k \mathbf{p}^{(k)})$$
$$= \mathbf{e}^{(k)} - \alpha_k \mathbf{A}\mathbf{p}^{(k)}. \tag{5.29}$$

Comparing the conjugate gradient algorithm to the steepest decent algorithm in Equation 5.22, the extra steps are to estimate the coefficient β_k in Equation 5.27 and to build the conjugate search direction $\mathbf{p}^{(k+1)}$ in Equation 5.25.

In practice, the conjugate gradient algorithm is often expressed in an alternative formula for the step length α_k as

$$\alpha_k = \frac{(\mathbf{e}^{(k)}, \mathbf{e}^{(k)})}{(\mathbf{p}^{(k)}, \mathbf{A}\mathbf{p}^{(k)})}, \tag{5.30}$$

and in an alternative formula for the conjugate coefficient β_k as

$$\beta_k = \frac{(\mathbf{e}^{(k+1)}, \mathbf{e}^{(k+1)})}{(\mathbf{e}^{(k)}, \mathbf{e}^{(k)})}. \tag{5.31}$$

These alternatives are proved in the Exercises. The implementation of the conjugate gradient algorithm may be summarised as follows.

We choose $\mathbf{x}^{(1)}$, which gives $\mathbf{e}^{(1)} = \mathbf{b} - \mathbf{A}\mathbf{x}^{(1)}$, and set $\mathbf{p}^{(1)} = \mathbf{e}^{(1)}$;
Then, iteration for $k = 1, 2, 3, \cdots$:

$$\alpha_k = \frac{(\mathbf{e}^{(k)}, \mathbf{e}^{(k)})}{(\mathbf{p}^{(k)}, \mathbf{A}\mathbf{p}^{(k)})},$$

$$\mathbf{x}^{(k+1)} = \mathbf{x}^{(k)} + \alpha_k \mathbf{p}^{(k)},$$

$$\mathbf{e}^{(k+1)} = \mathbf{e}^{(k)} - \alpha_k \mathbf{A}\mathbf{p}^{(k)}, \tag{5.32}$$

$$\beta_k = \frac{(\mathbf{e}^{(k+1)}, \mathbf{e}^{(k+1)})}{(\mathbf{e}^{(k)}, \mathbf{e}^{(k)})},$$

$$\mathbf{p}^{(k+1)} = \mathbf{e}^{(k+1)} + \beta_k \mathbf{p}^{(k)}.$$

From a computational point of view, note the simplicity of this conjugate gradient algorithm. It involves nothing more than

- the inner product of a matrix and a vector; and only one such product per iteration since calculated $\mathbf{A}\mathbf{p}^{(k)}$ can be stored;
- the inner product of two vectors; and
- the sum of a vector and a scalar timing a vector.

The conjugate gradient method is also applicable to weakly nonlinear problems, with an objective function which is presented in non-quadratic forms but is approximately quadratic near the minimum (Fletcher and Reeves, 1964). The gradient is numerically recalculated based on the current solution estimate. Minimisation for a non-quadratic function has a slower rate of convergence than that for a pure quadratic function. In both cases, either quadratic or non-quadratic, the performance can be enhanced by reverting the search direction periodically to the steepest descent direction, the negative gradient vector. That is, setting β_k to zero after a number of iterations and restarting.

5.4 Biconjugate gradient method

In the previous two sections, $\mathbf{A} = \mathbf{G}^T\mathbf{G}$ is a $N \times N$ symmetric matrix. When this square matrix $\mathbf{A}$ tends to zero, the inner product $(\mathbf{p}^{(k)}, \mathbf{A}\mathbf{p}^{(k)}) = (\mathbf{G}\mathbf{p}^{(k)}, \mathbf{G}\mathbf{p}^{(k)})$ tends to zero, and there will be significant rounding errors in calculating α_k and in turn $\mathbf{x}^{(k+1)}$, $\mathbf{e}^{(k+1)}$ and $\mathbf{p}^{(k+1)}$. In general, if $\mathbf{G}$ is near singular, $\mathbf{A} = \mathbf{G}^T\mathbf{G}$ will be severely ill-conditioned. For example, the following 2×2 matrix

$$\mathbf{G} = \begin{bmatrix} 1 & 1 \\ 1-\varepsilon & 1 \end{bmatrix}, \tag{5.33}$$

with $\varepsilon = 0.01$, has a condition number (the ratio of two eigenvalues) ≈ 400. However, the square matrix $\mathbf{A} = \mathbf{G}^T\mathbf{G}$ has a condition number $\approx 160,000$. For large-order sets of equations, there will be a strong risk of encountering poor convergence rates.

The biconjugate gradient method (Lanczos, 1950; Fletcher, 1976) can be expected to have convergence characteristics relating to the eigenvalues of $\mathbf{G}$ rather than $\mathbf{A} = \mathbf{G}^T\mathbf{G}$. It simultaneously solves the original problem and an additional yet unwanted problem:

$$\mathbf{G}\mathbf{x} = \mathbf{d} \quad \text{and} \quad \mathbf{G}^T\hat{\mathbf{x}} = \hat{\mathbf{d}}. \tag{5.34}$$

The objective function, which shall be optimised, is

$$\phi(\mathbf{e}, \hat{\mathbf{e}}) = (\hat{\mathbf{e}}, \mathbf{G}^{-1}\mathbf{e}), \tag{5.35}$$

where $\mathbf{e} = \mathbf{d} - \mathbf{G}\mathbf{x}$ and $\hat{\mathbf{e}} = \hat{\mathbf{d}} - \mathbf{G}^T\hat{\mathbf{x}}$. Starting initially with $\mathbf{p}^{(1)} = \mathbf{e}^{(1)}$ and $\hat{\mathbf{p}}^{(1)} = \hat{\mathbf{e}}^{(1)}$, the iterative procedure is

$$\alpha_k = \frac{(\hat{\mathbf{e}}^{(k)}, \mathbf{e}^{(k)})}{(\hat{\mathbf{p}}^{(k)}, \mathbf{Gp}^{(k)})},$$

$$\mathbf{x}^{(k+1)} = \mathbf{x}^{(k)} + \alpha_k \mathbf{p}^{(k)}, \qquad (5.36)$$

$$\mathbf{e}^{(k+1)} = \mathbf{e}^{(k)} - \alpha_k \mathbf{Gp}^{(k)},$$

$$\hat{\mathbf{e}}^{(k+1)} = \hat{\mathbf{e}}^{(k)} - \alpha_k \mathbf{G}\hat{\mathbf{p}}^{(k)}.$$

Note that the second solution update $\hat{\mathbf{x}}^{(k+1)} = \hat{\mathbf{x}}^{(k)} + \alpha_k \hat{\mathbf{p}}^{(k)}$ is not needed.

At each iteration, the update directions can be found by

$$\beta_k = \frac{(\hat{\mathbf{e}}^{(k+1)}, \mathbf{e}^{(k+1)})}{(\hat{\mathbf{e}}^{(k)}, \mathbf{e}^{(k)})},$$

$$\mathbf{p}^{(k+1)} = \mathbf{e}^{(k+1)} + \beta_k \mathbf{p}^{(k)}, \qquad (5.37)$$

$$\hat{\mathbf{p}}^{(k+1)} = \hat{\mathbf{e}}^{(k+1)} + \beta_k \hat{\mathbf{p}}^{(k)}.$$

Scalar α_k in Equation 5.36 is chosen so as to force the biorthogonality condition,

$$(\hat{\mathbf{e}}^{(k+1)}, \mathbf{e}^{(k)}) = (\mathbf{e}^{(k+1)}, \hat{\mathbf{e}}^{(k)}) = 0. \qquad (5.38)$$

Scalar β_k is chosen so as to force the biconjugacy condition

$$(\hat{\mathbf{p}}^{(k+1)}, \mathbf{Gp}^{(k)}) = (\mathbf{p}^{(k+1)}, \mathbf{G}\hat{\mathbf{p}}^{(k)}) = 0. \qquad (5.39)$$

In general, the biconjugate gradient algorithm holds the following properties:

$$(\hat{\mathbf{p}}^{(i)}, \mathbf{Gp}^{(j)}) = 0, \qquad \text{for } i \neq j,$$

$$(\hat{\mathbf{e}}^{(i)}, \mathbf{e}^{(j)}) = 0, \qquad \text{for } i \neq j, \qquad (5.40)$$

$$(\hat{\mathbf{e}}^{(i)}, \mathbf{p}^{(j)}) = (\mathbf{e}^{(i)}, \hat{\mathbf{p}}^{(j)}) = 0, \qquad \text{for } i > j.$$

The biconjugate gradient method can be easily extended for solving a complex-valued system (Appendix C). However, since there is no minimisation procedure within any intermediate steps, this method often shows irregular convergence behaviour. In some cases, rounding errors can even result in severe cancellation effects in the solution. If we minimise the residual explicitly within each iteration, the procedure would have a robust convergence. This is a stabilised version of the biconjugate gradient method (van der Vorst, 1992), as summarised in what follows.

Given an initial solution $\mathbf{x}^{(1)}$, we compute the initial residual $\mathbf{e}^{(1)} = \mathbf{d} - \mathbf{G}\mathbf{x}^{(1)}$; we choose $\hat{\mathbf{e}}^{(1)} = \mathbf{e}^{(1)}$, so that $(\hat{\mathbf{e}}^{(1)}, \mathbf{e}^{(1)}) \neq 0$; and then we set the initial search direction $\mathbf{p}^{(1)} = \mathbf{e}^{(1)}$.

Iterations are performed for $k = 1, 2, \cdots$:

1) Calculate the first half-step length parameter,

$$\alpha_k = \frac{(\hat{\mathbf{e}}^{(1)}, \mathbf{e}^{(k)})}{(\hat{\mathbf{e}}^{(1)}, \mathbf{G}\mathbf{p}^{(k)})} \, , \tag{5.41}$$

and compute the current solution and the residual

$$\mathbf{x}^{(k+1/2)} = \mathbf{x}^{(k)} + \alpha_k \, \mathbf{p}^{(k)} \, ,$$
$$\mathbf{e}^{(k+1/2)} = \mathbf{e}^{(k)} - \alpha_k \, \mathbf{G}\mathbf{p}^{(k)} \, . \tag{5.42}$$

2) Calculate a second half-step length parameter,

$$\omega_k = \frac{(\mathbf{G}\mathbf{e}^{(k+1/2)}, \mathbf{e}^{(k+1/2)})}{(\mathbf{G}\mathbf{e}^{(k+1/2)}, \mathbf{G}\mathbf{e}^{(k+1/2)})} \, , \tag{5.43}$$

and update the solution and the residual to

$$\mathbf{x}^{(k+1)} = \mathbf{x}^{(k+1/2)} + \omega_k \, \mathbf{e}^{(k+1/2)},$$
$$\mathbf{e}^{(k+1)} = (\mathbf{I} - \omega_k \, \mathbf{G})\mathbf{e}^{(k+1/2)}. \tag{5.44}$$

3) Find the biconjugate coefficient

$$\beta_k = \frac{\alpha_k}{\omega_k} \frac{(\hat{\mathbf{e}}^{(1)}, \mathbf{e}^{(k+1)})}{(\hat{\mathbf{e}}^{(1)}, \mathbf{e}^{(k)})} \, , \tag{5.45}$$

and compute the search direction

$$\mathbf{p}^{(k+1)} = \mathbf{e}^{(k+1)} + \beta_k(\mathbf{I} - \omega_k \, \mathbf{G}) \, \mathbf{p}^{(k)} \, . \tag{5.46}$$

Within each iteration, the first half-step length parameter α_k is obtained from the orthogonality between $\hat{\mathbf{e}}^{(1)} = \mathbf{e}^{(1)}$ and updated residual $\mathbf{e}^{(k+1/2)}$. It is easy to verify the inner product $(\hat{\mathbf{e}}^{(1)}, \mathbf{e}^{(k+1/2)}) = 0$. The second half-step length parameter ω_k is obtained by minimising the residual, $\min \| \mathbf{e}^{(k+1)}(\omega_k) \|^2 = \min \| (\mathbf{I} - \omega_k \, \mathbf{G})\mathbf{e}^{(k+1/2)} \|^2$. Since the residual at each iteration is minimised explicitly in a least-squares sense, this method converges in a smoother and faster way than the standard biconjugate gradient algorithm.

5.5 The subspace gradient method

Both the steepest descent and the conjugate gradient methods have ignored differences between different parameter types. Where the model depends on parameters of different dimensionality, applying a single step length to all parameters can result in very slow convergence. The subspace method (Kennett *et al.*, 1988) is ideally suited to problems in which the model space includes parameters of different dimensionality, such as velocity and reflector depth in seismic inversion.

Let us define an objective function by the data misfit:

$$\phi(\mathbf{x}) = [\mathbf{d} - G(\mathbf{x})]^T \mathbf{C}_d^{-1} [\mathbf{d} - G(\mathbf{x})] , \qquad (5.47)$$

where $\mathbf{d}$ is the observed data vector, $G(\mathbf{x})$ is the forward calculation prediction, $\mathbf{x} = \mathbf{m}$ is the vector of unknown parameters that describe the properties of the underground medium, and $\mathbf{C}_d$ is the data covariance matrix.

Assuming that $\phi(\mathbf{x})$ is a smooth function, we can make a locally quadratic approximation, about a current solution $\mathbf{x}$, using a truncated Taylor series for $\phi(\mathbf{x})$:

$$\phi(\mathbf{x} + \Delta\mathbf{x}) \approx \phi(\mathbf{x}) + \Delta\mathbf{x}^T \boldsymbol{\gamma} + \frac{1}{2}\Delta\mathbf{x}^T \mathbf{H}\, \Delta\mathbf{x} , \qquad (5.48)$$

where $\boldsymbol{\gamma}$ is the gradient vector, and $\mathbf{H}$ is the Hessian matrix.

The gradient of the objective function is defined by

$$\boldsymbol{\gamma} = \nabla_x \phi(\mathbf{x}) = -2\mathbf{F}^T \mathbf{C}_d^{-1}[\mathbf{d} - G(\mathbf{x})] , \qquad (5.49)$$

where

$$\mathbf{F} = \nabla_x G(\mathbf{x}) \qquad (5.50)$$

is the matrix of the Fréchet derivatives of the geophysical data $G(\mathbf{x})$, with respect to the model parameters $\mathbf{x}$.

The Hessian matrix $\mathbf{H}$ of the objective function can be calculated by

$$\mathbf{H} \equiv \nabla_x \nabla_x \phi(\mathbf{x}) = 2\mathbf{F}^T \mathbf{C}_d^{-1}\mathbf{F} + 2\nabla_x \mathbf{F}^T \mathbf{C}_d^{-1}[\mathbf{d} - G(\mathbf{x})] . \qquad (5.51)$$

Since $\nabla_x \mathbf{F} = \nabla_x \nabla_x \phi(\mathbf{x})$ appears with the data misfit, its significance should diminish as minimisation proceeds. Thus, the last term is often neglected at the outset.

Introducing $\mathbf{C}_x$, a model covariance matrix with unit (*model parameter*)^2, we get the steepest ascent vector $\hat{\boldsymbol{\gamma}}$ in the model space, in terms of the gradient vector :

$$\hat{\boldsymbol{\gamma}} = \mathbf{C}_x \boldsymbol{\gamma} . \tag{5.52}$$

The steepest descent method updates the solution estimate $\mathbf{x}$ along the steepest descent direction $(-\hat{\boldsymbol{\gamma}})$ with a step length that minimises $\phi(\mathbf{x})$. The subspace method, however, will partition the steepest ascent vector $\hat{\boldsymbol{\gamma}}$ into several independent subvectors and choose an optimal step length for each of them.

Introducing q basis vectors $\{\mathbf{a}^{(j)}\}$, where q is a relatively small number of subspaces, we compose a projection matrix $\mathbf{A}$ as

$$A_{ij} = a_i^{(j)} , \qquad \text{for } \begin{cases} i = 1, \cdots, N, \\ j = 1, \cdots, q, \end{cases} \tag{5.53}$$

where N is the length of the basis vectors, equal to the length of the model vector $\mathbf{x}$. Then, we construct the following perturbation to the current model in the space spanned by the $\{\mathbf{a}^{(j)}\}$:

$$\Delta \mathbf{x} = \sum_{j=1}^{q} \alpha_j \mathbf{a}^{(j)} \equiv \mathbf{A}\boldsymbol{\alpha} , \tag{5.54}$$

where the coefficients $\{\alpha_j, \text{ for } j = 1, \cdots, q\}$ are determined by minimising $\phi(\mathbf{x})$ for this class of perturbation. We can write Equation 5.48 in terms of the coefficient vector $\boldsymbol{\alpha}$, as

$$\phi(\mathbf{x} + \mathbf{A}\boldsymbol{\alpha}) \approx \phi(\mathbf{x}) + \boldsymbol{\alpha}^T \mathbf{A}^T \boldsymbol{\gamma} + \frac{1}{2} \boldsymbol{\alpha}^T \mathbf{A}^T \mathbf{H} \, \mathbf{A}\boldsymbol{\alpha} , \tag{5.55}$$

set $\partial \phi / \partial \boldsymbol{\alpha} = \mathbf{0}$, and finally determine the coefficient vector,

$$\boldsymbol{\alpha} = -[\mathbf{A}^T \mathbf{H} \mathbf{A} + \mu \mathbf{I}]^{-1} \mathbf{A}^T \boldsymbol{\gamma} , \tag{5.56}$$

where μ is a stabilisation factor. Then, we have the the following model update,

$$\Delta \mathbf{x} = -\mathbf{A} \, [\mathbf{A}^T \mathbf{H} \mathbf{A} + \mu \mathbf{I}]^{-1} \mathbf{A}^T \boldsymbol{\gamma} . \tag{5.57}$$

The subspace method is a very effective approach in which only the $q \times q$ matrix $[\mathbf{A}^T \mathbf{H} \mathbf{A} + \mu \mathbf{I}]$ needs to be inverted. This small $q \times q$ projected

Hessian matrix is generally well conditioned with judicious choices for the basis vectors $\{\mathbf{a}^{(j)}\}$.

The basis vectors can be related to the steepest ascent vector $\hat{\boldsymbol{\gamma}}$. Assume that the model parameters are classified into several different parameter types, say $\mathbf{x}_I$, for $I = A, B, \cdots$. Concentrating on one class of model parameters at a time, the gradient component is

$$\boldsymbol{\gamma}_I = \nabla_{\mathbf{x}_I} \phi(\mathbf{x}), \qquad \text{for } I = A, B, \cdots \tag{5.58}$$

We construct the corresponding steepest ascent vector in full model space:

$$\hat{\boldsymbol{\gamma}}_A = \mathbf{C}_\mathbf{x} \begin{bmatrix} \boldsymbol{\gamma}_A \\ 0 \\ \vdots \\ \vdots \\ 0 \end{bmatrix}, \quad \hat{\boldsymbol{\gamma}}_B = \mathbf{C}_\mathbf{x} \begin{bmatrix} 0 \\ \boldsymbol{\gamma}_B \\ 0 \\ \vdots \\ 0 \end{bmatrix}, \quad \cdots \tag{5.59}$$

The $\hat{\boldsymbol{\gamma}}_I$ are projection vectors of the gradient components $\boldsymbol{\gamma}_I$, for each parameter type. Then we build the basis vectors $\{\mathbf{a}^{(j)}\}$ as

$$\mathbf{a}^{(1)} = \frac{1}{\|\hat{\boldsymbol{\gamma}}_A\|} \hat{\boldsymbol{\gamma}}_A, \quad \mathbf{a}^{(2)} = \frac{1}{\|\hat{\boldsymbol{\gamma}}_B\|} \hat{\boldsymbol{\gamma}}_B, \quad \cdots \tag{5.60}$$

If we ignore the projection matrix $\mathbf{A}$, the subspace gradient method of Equation 5.57 becomes

$$\Delta \mathbf{x} = -[\mathbf{H} + \mu \mathbf{I}]^{-1} \boldsymbol{\gamma}. \tag{5.61}$$

Note here the step length is defined by the inverse of the Hessian matrix. We will discuss this case in the next chapter.

CHAPTER 6

Regularisation

In an inverse problem, the quality of data fitting between theoretical and experimental observation can be measured using a χ^2 (chi-square) test. Minimising the data misfit is equivalent to maximising a likelihood function, and the latter is a synonym for data probability.

However, seismic inversion is inherently ill-posed, and needs proper regularisation for producing a well-behaved model estimate. From the statistical point of view, the minimisation problem with a model constraint is equivalent to maximising a conditional probability, given the data measurements. In other words, any probability function that can appropriately describe the behaviour of the expected model can be adopted as a regularisation in the inversion.

6.1 Regularisation versus conditional probability

Suppose we have a set of measurements in vector $\tilde{\mathbf{d}}$. Each measurement $\tilde{d}_i$ has its error e_i. Assuming that the errors $\{e_i\}$ happen to be uncorrelated and the mean is zero, the χ^2 test is defined as

$$\chi^2 = \sum_{i=1}^{M} \frac{e_i^2}{\sigma_i^2}, \tag{6.1}$$

where M is the total number of data samples, and σ_i^2 is the variance for the ith sample.

In seismic inversion, this χ^2 test can be used to measure how well the model estimate $\mathbf{x}$ agrees with the observation $\tilde{\mathbf{d}}$. As the data residual

Seismic Inversion: Theory and Applications, First Edition. Yanghua Wang.
© 2017 John Wiley & Sons, Ltd. Published 2017 by John Wiley & Sons, Ltd.

corresponding to a discrete model $\mathbf{x}$ is $\mathbf{e} = \tilde{\mathbf{d}} - G(\mathbf{x})$, then the χ^2 test can be written as

$$\chi^2 = [\tilde{\mathbf{d}} - G(\mathbf{x})]^T \mathbf{C}_d^{-1} [\tilde{\mathbf{d}} - G(\mathbf{x})] , \tag{6.2}$$

where $\mathbf{C}_d$ is the data covariance matrix, in which the nonzero main diagonal elements are $C_{ii} = \sigma_i^2$. The off-diagonal covariance can be nonzero valued, and thus Equation 6.2 is a generalised version of the χ^2 test. The inverse problem of data fitting is to minimise the χ^2 distribution of data residuals, for producing a χ^2 quasi-solution: $\phi(\mathbf{x}) = \chi^2$.

The χ^2 test may be related to the Gaussian distribution of data errors as follows. For a single data error e_i, if it is an independent variable, the Gaussian distribution is defined as

$$p(e_i) = \frac{1}{\sqrt{2\pi}\sigma_i} \exp\left(-\frac{e_i^2}{2\sigma_i^2} \right). \tag{6.3}$$

If $\{e_i\}$ are independent standard normal variables, with probability functions $\{p(e_i)\}$, the joint probability is just the product of these univariate distributions:

$$p(\mathbf{e}) = p(e_1)p(e_2)\cdots p(e_M) . \tag{6.4}$$

Since $\{e_i\}$ has a χ^2 distribution, the joint probability is

$$p(\mathbf{e}) = \frac{1}{(2\pi)^{M/2} \prod\limits_{i=1}^{M} \sigma_i} \exp\left(-\frac{1}{2}\chi^2 \right). \tag{6.5}$$

Therefore, the inverse problem for minimising the χ^2 test is equivalent to maximising the probability $p(\mathbf{e})$.

Considering that data vector $\tilde{\mathbf{d}}$ is a fixed parameter in this probability function, model vector $\mathbf{x}$ is the variable that is allowed to vary freely. From this point of view, the distribution function can also be called the likelihood function:

$$\ell(\mathbf{x} \mid \tilde{\mathbf{d}}) = p(\tilde{\mathbf{d}} \mid \mathbf{x}), \tag{6.6}$$

where $p(\tilde{\mathbf{d}} \mid \mathbf{x}) = p(\mathbf{e})$ as $\mathbf{e} = \tilde{\mathbf{d}} - G(\mathbf{x})$. Maximising the likelihood function means to reconstruct a model $\mathbf{x}$, equivalent to constraining the distribution of model parameters, in order to make the data errors $\tilde{\mathbf{d}} - G(\mathbf{x})$ have a maximum probability.

Therefore, the following three concepts have exactly the same functionality: maximising the likelihood of model distribution, maximising the distribution of data errors, and minimising the χ^2 measurement of data errors.

In the χ^2 quasi-solution of Equation 6.2, if we neglect the off-diagonal covariance, we can present it as

$$\phi(\mathbf{x}) = [\tilde{\mathbf{d}} - G(\mathbf{x})]^{\mathrm{T}} \mathbf{W} [\tilde{\mathbf{d}} - G(\mathbf{x})], \tag{6.7}$$

where $\mathbf{W}$ is a diagonal weighting matrix with weights $w_{ii} = \sigma_i^{-2}$. This is a weighted least-squares problem. Practically, the diagonal weighting matrix reflects our confidence on each observation sample. For an accurate observation that has a small variance, a great weight is assigned to the prediction errors e_i in the objective function. For an inaccurate observation with a great variance, a small weight is assigned to the prediction errors e_i.

In the case of uncorrelated data with uniform variance, that is, $\mathbf{W} = \sigma^{-2} \mathbf{I}$, then Equation 6.7 is the least-squares problem presented in the previous chapters, $\phi(\mathbf{x}) = \| \tilde{\mathbf{d}} - G(\mathbf{x}) \|^2$. Therefore, we can conclude that the least-squares problem is to maximise the likelihood function defined as a Gaussian probability.

The inverse problem of data fitting in the least-squares sense may need additional regularisation, which allows *a priori* information to be incorporated within the solution estimation. For example, an L_2 norm constraint for a smooth solution, an L_1 norm constraint for a spike inversion, an maximum entropy constraint for sharpening an image, and the Cauchy constraint for sparse inversion.

The least-squares data fitting with an additive model constraint can be presented as a maximisation to a conditional probability, in which regularisation becomes a prior probability of model $\mathbf{x}$, being applied jointly with the data probability.

The desired conditional probability of 'model' vector $\mathbf{x}$, for given 'data' vector $\tilde{\mathbf{d}}$, is

$$p(\mathbf{x} \mid \tilde{\mathbf{d}}) = \frac{p(\tilde{\mathbf{d}} \mid \mathbf{x}) p(\mathbf{x})}{p(\tilde{\mathbf{d}})}, \tag{6.8}$$

where $p(\tilde{\mathbf{d}} \mid \mathbf{x})$ is the conditional probability of the observation $\tilde{\mathbf{d}}$ with the parameter $\mathbf{x}$, $p(\mathbf{x})$ is the a prior probability of model $\mathbf{x}$, and $p(\tilde{\mathbf{d}})$ is the probability of observation. This is Bayes' theorem.

Following Bayes' theorem, the conditional probability $p(\mathbf{x} \mid \tilde{\mathbf{d}})$ is equal to the joint probability, $p(\tilde{\mathbf{d}} \mid \mathbf{x})p(\mathbf{x})$, divided by the marginal probability $p(\tilde{\mathbf{d}})$. The latter is a constant

$$p(\tilde{\mathbf{d}}) = \int p(\tilde{\mathbf{d}} \mid x)p(x)\mathrm{d}x. \tag{6.9}$$

Therefore, we have

$$\max p(\mathbf{x} \mid \tilde{\mathbf{d}}) = \frac{1}{p(\tilde{\mathbf{d}})} \max p(\tilde{\mathbf{d}} \mid \mathbf{x})p(\mathbf{x}) . \tag{6.10}$$

Note that $p(\mathbf{x} \mid \tilde{\mathbf{d}})$ is really only a probability of the model $\mathbf{x}$, with the data $\tilde{\mathbf{d}}$ just providing auxiliary information. This method with a maximum posterior distribution is called the maximum *a posteriori* probability (MAP) estimation.

MAP is different from the maximum likelihood. According to Equation 6.6, the maximum likelihood, $\max \ell(\mathbf{x} \mid \tilde{\mathbf{d}})$, is equivalent to maximising $p(\tilde{\mathbf{d}} \mid \mathbf{x})$. The MAP in Equation 6.9, $\max p(\mathbf{x} \mid \tilde{\mathbf{d}})$, means to maximise $p(\tilde{\mathbf{d}} \mid \mathbf{x})p(\mathbf{x})$, which incorporates $p(\tilde{\mathbf{d}} \mid \mathbf{x})$ with a prior distribution $p(\mathbf{x})$. As $p(\mathbf{x})$ is closely related to the regularisation, MAP may avoid potential over-fitting by the maximum likelihood. MAP is closely related to the maximum likelihood, and tends to look like the maximum likelihood, asymptotically.

In summary, least-squares inversion of data fitting can be interpreted as maximising the probability of data or errors with a Gaussian distribution, and the least-squares inversion plus various regularisation can be interpreted as maximising the posterior distribution of the model solution.

6.2 The L_p norm constraint

In seismic inversion, the general L_p norm with $0 < p \leq 2$ is often used to define a model constraint. The L_p norm of a vector is defined by

$$L_p = \| \mathbf{x} \|_p = \left(\sum_k | x_k |^p \right)^{1/p} , \tag{6.11}$$

where 'L' is named after a French mathematician, Henri Léon Lebesgue.

The physical meaning of the vector norm is a measure of distance. For the three popular norms, the physical meanings are straightforward.

1) The L_1 norm $\| \mathbf{x} \|_1$ is the sum of the absolute values of its components. Thus, it is a measurement of so-called taxicab geometry, or so-called city block distance in a rectangular street grid.

2) The L_2 norm $\| \mathbf{x} \|_2$ is the square root of the sum of squares of its components. It is the ordinary distance from the origin to the point $\mathbf{x}$.

3) The L_∞ norm $\| \mathbf{x} \|_\infty$ is the maximum of the absolute values. Therefore, it is the longest street the taxi drives through.

If $\mathbf{x}$ is a complex vector, $|x_k|$ in the definition of Equation 6.11 are the modules of the complex numbers x_k. This choice ensures the norm of vector $\mathbf{x}$ is a non-negative real number.

In seismic inversion, for convenience, we do not use the definition of L_p norm in Equation 6.11 and, instead, use $\| \mathbf{x} \|_p^p$ as a constraint:

$$\| \mathbf{x} \|_p^p = \sum_k |x_k|^p . \tag{6.12}$$

Hence, the objective function is given by

$$\phi(\mathbf{x}) = \| \tilde{\mathbf{d}} - G(\mathbf{x}) \|_2^2 + \mu \| \mathbf{x} - \mathbf{x}_{\text{ref}} \|_p^p . \tag{6.13}$$

Theoretically, the goodness of data fitting can also be measured with a general L_p norm, for example, $p = 1$ (Crase *et al.*, 1990; Brossier *et al.*, 2010). But we focus our discussion on the L_p norm model constraint.

First, let us compare the L_1 norm and the L_2 norm. Since the L_1 norm is the sum of absolute sample values, and the L_2 norm is the sum of squared sample values, the L_1 norm weights the *bad* samples (outliers) less than the L_2 norm, in which the square operation makes a big value even large in contrast to the rest. In other words, the inversion with the L_1 norm constraint allows the existence of some outliers in the solution vector $\mathbf{x}$. The inversion with the L_2 norm constraint has much weak tolerance. As the inversion with the L_∞ norm constraint has the lowest tolerance, we can say that the tolerance to any outliers is inversely proportional to the p value.

Second, the minimisation requires calculation of the first-order derivative of the L_p norm term,

$$\frac{\partial \| \mathbf{x} \|_p^p}{\partial x_k} = p x_k |x_k|^{p-2} , \tag{6.14}$$

where x_k is an element of vector $\mathbf{x}$. For $p = 1$ and $p = 2$ we have

$$\frac{\partial \| \mathbf{x} \|_{p=1}}{\partial x_k} = \text{sgn}(x_k), \qquad \frac{\partial \| \mathbf{x} \|_{p=2}^2}{\partial x_k} = 2x_k. \tag{6.15}$$

where $\text{sgn}(x_k)$ is the sign of x_k. Consider that each of these is an additive part of the gradient vector γ, and model update $\Delta \mathbf{x}$ follows the (negative) gradient direction, with a constant step length α. The model constraint makes a contribution to the gradient component $\partial \phi / \partial x_k$ in the following ways: the L_1 norm constraint adds on the sign of x_k, that is, either $+1$ or -1, and the L_2 norm constraint adds on the current model value x_k. Since the L_1 norm constraint takes the sign and neglects its magnitude, it will make the model solution to be more spiked than the L_2 norm constraint.

Finally, let us find the solution of the L_p norm inversion. Assuming $G(\mathbf{x}) = \mathbf{Gx}$ and $\mathbf{x}_{\text{ref}} = \mathbf{0}$, the objective function is defined as

$$\phi(\mathbf{x}) = (\tilde{\mathbf{d}} - \mathbf{Gx})^{\text{T}} \mathbf{C}_d^{-1} (\tilde{\mathbf{d}} - \mathbf{Gx}) + \mu \| \mathbf{x} \|_p^p. \tag{6.16}$$

Setting $\partial \phi / \partial \mathbf{x} = \mathbf{0}$, we can obtain the solution:

$$\mathbf{x} = [\mathbf{G}^{\text{T}} \mathbf{C}_d^{-1} \mathbf{G} + \mu \mathbf{D}]^{-1} \mathbf{G}^{\text{T}} \mathbf{C}_d^{-1} \tilde{\mathbf{d}}, \tag{6.17}$$

where $\mathbf{D}$ is a diagonal matrix:

$$\mathbf{D} = \frac{p}{2} \text{diag}\{ | x_1 |^{p-2}, | x_2 |^{p-2}, \cdots, | x_N |^{p-2} \}, \tag{6.18}$$

depending on the current estimate. For the calculation of this diagonal matrix, we propose to make a stabilisation as in the following:

$$| x_k |^{p-2} \approx \frac{| x_k |^p + \sigma^2}{x_k^2 + \sigma^2}, \tag{6.19}$$

where σ^2 is a small positive stabilisation factor. In contrast to a conventional stabilisation, σ^2 here is added not only to the denominator but also to the numerator (Wang, 2004a, 2006). When $x_k^2 \to 0$, the conventional stabilisation makes $| x_k |^{p-2} \to 0$, but the scheme of Equation 6.19 makes $| x_k |^{p-2} \to 1$. This is a general solution applicable to $0 < p \leq 2$. Therefore, we have a unified form for the least-squares solution, the L_1 norm solution and the solution for the L_p norm, with p less than 2.

6.3 The maximum entropy constraint

Physically, *entropy* is a measure of the disorder of a system. In the information theory, entropy is a measure of the uncertainty associated with a random variable (Jaynes, 1968). However, in geophysical inversion, the maximum entropy criterion has nothing to do with physical entropy. It is just a regularisation terminology used to constrain the inverse problem, in which we attempt to abstract most information from an erroneous data set, by maximising the entropy of model variables in an inversion solution.

The entropy of a discrete set of model samples $\{x_1, x_2, \cdots, x_N\}$ is defined as

$$E(\mathbf{x}) = -\sum_{k=1}^{N} |x_k| \ln |x_k|. \tag{6.20}$$

We are seeking a generalised solution $\mathbf{x}$ with a maximised $E(\mathbf{x})$, which equivalently means to minimise the negative entropy, $-E(\mathbf{x})$. The latter is the certainty measure of random variables in $\mathbf{x}$. Therefore, one can define the following objective function that will be minimised in the inversion as

$$\phi(\mathbf{x}) = [\tilde{\mathbf{d}} - G(\mathbf{x})]^{\mathrm{T}} \mathbf{C}_{\mathrm{d}}^{-1} [\tilde{\mathbf{d}} - G(\mathbf{x})] + \mu \sum_{k=1}^{N} |x_k| \ln |x_k|, \tag{6.21}$$

where the trade-off parameter μ balances the data fitting and the minimum negative entropy regularisation. As the minimisation of the negative entropy is equivalent to the maximisation of the entropy, it is conventionally referred to as the maximum entropy method.

Compared to the L_2 norm constraint $(\mathbf{x}, \mathbf{x}) = \sum_k x_k^2$, in the maximum entropy constraint $\min \sum_k |x_k| \ln |x_k|$, the logarithmic operation boosts the details of low-intensity ripples. Including the logarithmic weighting in a minimisation problem will suppress the low-intensity ripples and sharpen the point events. Therefore, the maximum entropy method might endow images with better resolution in some cases.

The maximum entropy constraint, $\min \sum_k |x_k| \ln |x_k|$, is not a quadratic function of $\mathbf{x}$. Let us assume that $G(\mathbf{x}) = \mathbf{Gx}$, the objective function can be expressed as

$$\phi(\mathbf{x}) = (\tilde{\mathbf{d}} - \mathbf{Gx})^{\mathrm{T}} \mathbf{C}_{\mathrm{d}}^{-1} (\tilde{\mathbf{d}} - \mathbf{Gx}) + \mu \sum_{k=1}^{N} |x_k| \ln |x_k|. \tag{6.22}$$

Applying the minimal variation principle to this objective function leads to a nonlinear system. To solve this nonlinear problem, the following gradient method may be used.

The second order derivative of the objective function, with respect to a specific variable x_k, can be approximated as

$$\frac{\partial^2 \phi}{\partial x_k^2} = \frac{1}{\Delta x_k} \left(\frac{\partial \phi}{\partial x} \bigg|_{x_k + \Delta x_k} - \frac{\partial \phi}{\partial x} \bigg|_{x_k} \right). \tag{6.23}$$

Minimisation means attempting to make the gradient $\partial \phi / \partial x$ vanish at $x_k + \Delta x_k$. Therefore, we obtain a stabilised model update:

$$\Delta \mathbf{x} = - \left(\frac{\partial^2 \phi}{\partial \mathbf{x}^2} + \mu \mathbf{I} \right)^{-1} \frac{\partial \phi}{\partial \mathbf{x}}, \tag{6.24}$$

where μ is a stabilisation factor. The model can be updated along the negative gradient direction with a step length that is inversely proportional to the second order derivatives. It may be presented in vector-matrix form as

$$\Delta \mathbf{x} = -[\mathbf{H} + \mu \mathbf{I}]^{-1} \boldsymbol{\gamma}, \tag{6.25}$$

where $\boldsymbol{\gamma} \equiv \partial \phi / \partial \mathbf{x}$ is the gradient, and $\mathbf{H} \equiv \partial^2 \phi / \partial \mathbf{x}^2$ is the Hessian matrix of the second order derivatives of the objective function. Equation 6.25 is identical to Equation 5.61 presented in the previous chapter.

To better understand the solution in Equation 6.25, let us revisit the minimisation problem $\phi(\mathbf{x}) = \| \mathbf{d} - \mathbf{Gx} \|^2$. The gradient vector is

$$\boldsymbol{\gamma} \equiv \frac{\partial \phi}{\partial \mathbf{x}} = -2\mathbf{G}^{\mathrm{T}}(\mathbf{d} - \mathbf{Gx}) = -2\mathbf{G}^{\mathrm{T}}\mathbf{e}, \tag{6.26}$$

and the Hessian matrix is

$$\mathbf{H} \equiv \frac{\partial^2 \phi}{\partial \mathbf{x}^2} \approx 2\mathbf{G}^{\mathrm{T}}\mathbf{G}. \tag{6.27}$$

Hence, the solution for this problem is

$$\Delta \mathbf{x} = [\mathbf{G}^{\mathrm{T}}\mathbf{G} + \mu \mathbf{I}]^{-1}\mathbf{G}^{\mathrm{T}}\mathbf{e}. \tag{6.28}$$

When the inverse matrix $[\mathbf{G}^{\mathrm{T}}\mathbf{G} + \mu\mathbf{I}]^{-1}$ is difficult to calculate, an approximation to $[\mathbf{G}^{\mathrm{T}}\mathbf{G}]$ is used. The simplest approximation is a diagonal matrix made by the diagonal elements of $[\mathbf{G}^{\mathrm{T}}\mathbf{G}]$. This is the classical Jacobi's method. We can adopt the same strategy to solve the maximum entropy problem as in what follows.

Denote the two terms, the quality and the regularisation in the objective function of Equation 6.22, as

$$Q = (\tilde{\mathbf{d}} - \mathbf{G}\mathbf{x})^{\mathrm{T}}\mathbf{C}_{\mathrm{d}}^{-1}(\tilde{\mathbf{d}} - \mathbf{G}\mathbf{x}),\tag{6.29}$$

$$R = \sum_{k=1}^{N} |\mathbf{x}_k| \ln |x_k| = f(\mathbf{x}).\tag{6.30}$$

The gradients and the second-order partial derivative matrices can be computed as

$$\frac{\partial Q}{\partial \mathbf{x}} = -2\mathbf{G}^{\mathrm{T}}\mathbf{C}_{\mathrm{d}}^{-1}(\mathbf{d} - \mathbf{G}\mathbf{x}),\qquad \frac{\partial^2 Q}{\partial \mathbf{x}^2} = 2\mathbf{G}^{\mathrm{T}}\mathbf{C}_{\mathrm{d}}^{-1}\mathbf{G},\tag{6.31}$$

and

$$\frac{\partial R}{\partial \mathbf{x}} = f'(\mathbf{x}),\qquad \frac{\partial^2 R}{\partial \mathbf{x}^2} = f''(\mathbf{x}).\tag{6.32}$$

Notice that the second order derivative of R is a diagonal matrix. In an iterative procedure, the second order derivative of Q can also be approximated to a diagonal matrix. Therefore, the Hessian matrix is approximated by a diagonal matrix $\mathbf{D}$ with the element defined by

$$D_{kk} = \left(2\mathbf{G}^{\mathrm{T}}\mathbf{C}_{\mathrm{d}}^{-1}\mathbf{G} + \frac{\partial^2 R}{\partial \mathbf{x}^2}\right)_{kk},\tag{6.33}$$

where $(\mathbf{G}^{\mathrm{T}}\mathbf{G})_{kk}$ is a sum of the element squares along the kth column of matrix $\mathbf{G}$. Solution (6.25) becomes

$$\Delta \mathbf{x} = -[\mathbf{D} + \mu\mathbf{I}]^{-1}\left(\frac{\partial Q}{\partial \mathbf{x}} + \mu\frac{\partial R}{\partial \mathbf{x}}\right).\tag{6.34}$$

After modifying $\mathbf{x}$, we can go back to the objective function defined by Equation 6.22 and run the procedure iteratively.

6.4 The Cauchy constraint

In statistics, the Cauchy distribution describes a distribution of random angles. In Figure 6.1, if the height λ is fixed, that is, the rotation point of a line segment is fixed, the tilted line segment cuts the horizontal axis at distance x. The Cauchy distribution describes a continuous distribution of the horizontal distance x, corresponding to the random angle θ.

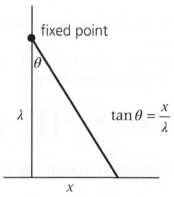

fixed point

λ

$\tan\theta = \dfrac{x}{\lambda}$

x

Figure 6.1 The Cauchy distribution describes the distribution of a random angle θ with respect to the vertical axis. The height is fixed to be λ, that is, the rotation point of a line segment is fixed, and the tilted line segment cuts the horizontal axis at distance x.

The random angle θ of the tilted line segment with respect to the vertical axis is $\theta = \tan^{-1}(x/\lambda)$, and the perturbation of the angle θ corresponding to the perturbation in the horizontal distance x is

$$d\theta = \frac{dx}{\lambda(1 + x^2/\lambda^2)}. \tag{6.35}$$

Hence, the distribution of angle θ is given by

$$\frac{d\theta}{\pi} = \frac{1}{\pi}\frac{\lambda}{(\lambda^2 + x^2)}dx. \tag{6.36}$$

This is normalised over all angles, since

$$\int_{-\pi/2}^{\pi/2} \frac{d\theta}{\pi} = 1, \tag{6.37}$$

$$\int_{-\infty}^{\infty} \frac{\lambda}{\pi(\lambda^2 + x^2)} dx = \frac{1}{\pi} \left[\tan^{-1}\left(\frac{x}{\lambda}\right) \right]_{-\infty}^{\infty} = 1. \tag{6.38}$$

Therefore, following Equation 6.36, the general Cauchy probability density function is defined as

$$p(x_k) = \frac{1}{\pi} \frac{\lambda}{\lambda^2 + x_k^2}, \tag{6.39}$$

where λ is the half width at half maximum. In this expression, x_k is assumed to have a zero mean.

For an N-dimensional event $\mathbf{x} = \{x_1, x_2, \cdots, x_N\}$, whose components are independent to each other, the joint probability is

$$p(\mathbf{x}) = p(x_1)p(x_2)\cdots p(x_N) = \prod_{k=1}^{N} \frac{1}{\pi} \frac{\lambda}{\lambda^2 + x_k^2}. \tag{6.40}$$

Using exponential and logarithmic expressions, we can transfer the product to become a summation:

$$p(\mathbf{x}) = \exp\left(-N \ln(\pi \lambda) - \sum_{k=1}^{N} \ln\left(1 + \frac{x_k^2}{\lambda^2} \right) \right). \tag{6.41}$$

Therefore, maximising the probability $p(\mathbf{x})$ leads to minimising the following function

$$\sum_{k=1}^{N} \ln\left(1 + \frac{x_k^2}{\lambda^2} \right). \tag{6.42}$$

This is the Cauchy constraint often used in the inverse problem.

Combining this Cauchy model constraint with the least-squares data fitting term, the objective function becomes

$$\phi(\mathbf{x}) = (\mathbf{d} - \mathbf{G}\mathbf{x})^{\mathrm{T}} \mathbf{C}_{\mathrm{d}}^{-1}(\mathbf{d} - \mathbf{G}\mathbf{x}) + \mu \sum_{k=1}^{N} \ln\left(1 + \frac{x_k^2}{\lambda^2} \right). \tag{6.43}$$

Setting $\partial\phi / \partial\mathbf{x} = \mathbf{0}$, we obtain the solution

$$\mathbf{x} = \left(\mathbf{G}^{\mathrm{T}} \mathbf{C}_{\mathrm{d}}^{-1} \mathbf{G} + \frac{\mu}{\lambda^2} \mathbf{D} \right)^{-1} \mathbf{G}^{\mathrm{T}} \mathbf{C}_{\mathrm{d}}^{-1} \mathbf{d}, \tag{6.44}$$

where $\mathbf{D}$ is a diagonal matrix defined as (Wang, 2003a)

$$\mathbf{D} = \text{diag}\left\{ \left(1+\frac{x_1^2}{\lambda^2}\right)^{-1}, \left(1+\frac{x_2^2}{\lambda^2}\right)^{-1}, \cdots, \left(1+\frac{x_N^2}{\lambda^2}\right)^{-1} \right\}. \tag{6.45}$$

Note that solution (6.44) is in the same form as the least-squares solution $\mathbf{x} = [\mathbf{G}^T\mathbf{C}_d^{-1}\mathbf{G} + \mu\mathbf{I}]^{-1}\mathbf{G}^T\mathbf{C}_d^{-1}\mathbf{d}$, where the identity matrix $\mathbf{I}$ is replaced with $\lambda^{-2}\mathbf{D}$.

In the Cauchy constrained inversion, the Cauchy parameter λ plays a key role in controlling the sparseness of the inversion result. Recall that the Cauchy constraint is related to the joint probability of Equation 6.41, $p(\mathbf{x}) = \exp[-R(\lambda)]$, where $R(\lambda)$ is the regularisation:

$$R(\lambda) = N\ln(\pi\lambda) + \sum_{k=1}^{N}\ln\left(1+\frac{x_k^2}{\lambda^2}\right). \tag{6.46}$$

Minimising $R(\lambda)$ by set $\partial R/\partial\lambda = 0$, we obtain the following equation:

$$\sum_{k=1}^{N}\frac{x_k^2}{\lambda^2+x_k^2} = \frac{N}{2}. \tag{6.47}$$

We can solve this equation numerically and find an optimum λ value theoretically for the Cauchy constraint.

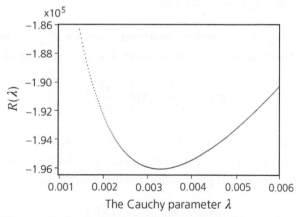

Figure 6.2 The regularisation function $R(\lambda)$ versus the Cauchy parameter λ. The estimated λ value corresponding to the minimum R value is $\lambda = 0.0033$.

As the inversion is performed iteratively, based on the statistical information in the current inversion model, we can estimate this sparseness parameter using Equation 6.47. Figure 6.2 displays a sample $R(\lambda)$ curve with the minima at $\lambda = 0.0033$.

Consider a case of seismic inversion retrieving reflectivity series from reflection seismic data. This reflectivity inversion is a pivotal step in seismic inversion, because the subsurface reflectivity model is directly linked with the vertical variation between different sedimentary sequences. It is appropriate to require the inverted model be sparse and the solution contain a few nonzero samples (reflectors). However, if we use the estimated λ value directly in the Cauchy inversion, the inversion result would be too sparse, and reflectivity samples with small amplitudes would be largely suppressed. Suppression of small seismic reflectivities would eliminate minor geological features. Thus, the estimated Cauchy parameter λ is not completely appropriate, and the reflectivity series does not strictly satisfy the Cauchy distribution.

Nevertheless, the estimated Cauchy parameter λ can be used as a good guide rather than just having a wild guess for the Cauchy parameter λ. We need to tune the trade-off parameter μ, for weakening the influence of the Cauchy constraint in the objective function, and make a compromise between the sparseness and the inversion residual.

6.5 Comparison of various regularisations

L_1 Norm constraint vs L_2 norm constraint

Minimising an L_2 norm model constraint, $\min \| \mathbf{x} \|_{p=2}^2$, is equivalent to maximising a Gaussian density distribution. The latter can be defined as

$$p(x) = \frac{1}{\sqrt{2\pi}\sigma} \exp\left(-\frac{x^2}{2\sigma^2} \right), \tag{6.48}$$

where σ^2 is the uniform model variance. Similarly, we can express the minimisation of an L_1 norm constraint, $\min \| \mathbf{x} \|_{p=1}$, equivalent to maximising an exponential probability density distribution:

$$p(x) = \frac{1}{2\sqrt{2}\sigma} \exp\left(-\frac{|x|}{\sqrt{2}\sigma} \right). \tag{6.49}$$

Figure 6.3 is a comparison between the Gaussian distribution and the exponential probability of equal variance $\sigma^2 = 1$. It shows that the exponential probability density function (solid curve) is much longer-tailed than the Gaussian function (dotted curve).

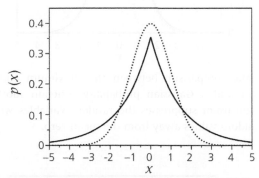

Figure 6.3 Schematic comparison between an exponential probability density function (solid curve) and a Gaussian probability density function (dotted curve). The exponential probability density function is longer-tailed than the Gaussian function.

The maximum entropy constraint vs L_2 norm constraint

The maximum entropy constraint, $\max E(\mathbf{x})$, is equivalent to a maximisation, $\max \exp(-\sum_k |x_k| \ln |x_k|)$. Hence, we compare the Gaussian function with the following exponential function:

$$p(x) = \frac{1}{4}\exp(-|x| \ln |x|). \tag{6.50}$$

Note that the entropy itself $E(\mathbf{x}) = -\sum_k |x_k| \ln |x_k|$ is a proper definition of the probability density function. But Equation 6.50 will lead to a joint probability function

$$p(\mathbf{x}) = \frac{1}{4}\exp\left(-\sum_{k=1}^{N} |x_k| \ln |x_k|\right). \tag{6.51}$$

Nevertheless, by convention, it is still called the maximum entropy method.

Figure 6.4 shows the so-called maximum entropy constraint (solid curve) suppressing the random variables with low absolute values and

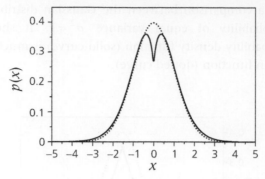

Figure 6.4 Schematic comparison between the maximum entropy probability function (solid curve) and a Gaussian probability function (dotted curve). The maximum entropy constraint suppresses the random variables with low absolute values and pushes random values away from the mean zero.

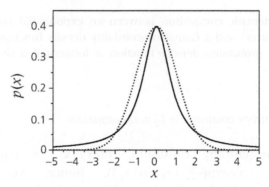

Figure 6.5 Schematic comparison between a Cauchy probability density function (solid curve) and a Gaussian probability density function (dotted curve).

pushing random values away from the mean zero. Except for this low absolute value part, the maximum entropy curve is close to the Gaussian distribution with $\sigma^2 = 1$ (dotted curve).

The Cauchy constraint vs L_2 norm constraint

The Cauchy probability density function is defined by Equation 6.39, as

$$p(x) = \frac{1}{\pi} \frac{\lambda}{\lambda^2 + x^2}. \tag{6.52}$$

Figure 6.5 is a schematic comparison regarding the probability density functions between the Gaussian distribution and the Cauchy distribution. For the sake of comparison, the Cauchy parameter λ is set as

$$\lambda = \sqrt{\frac{2}{\pi}}\sigma \approx 0.8\sigma \,, \tag{6.53}$$

so that both have the same maximum value. In Figure 6.5, $\sigma = 1$ and $\lambda = 0.8$ are used. It indicates that the Cauchy probability density function (solid curve) has a symmetric 'bell shaped' distribution, but is more peaked at the centre and has longer and fatter tails than a normal Gaussian distribution (dotted curve). In this aspect, the Cauchy constraint and L_1 norm constraint are quite similar as both have fat tails. When applying them as a model constraint, the inversion allows the resultant image have variables with large values.

In summary, regularisation can be a model constraint used together with the data fitting in an objective function. It also can be applied directly to the geophysical operator, as a method shown in the following chapter.

CHAPTER 7

Localised average solutions

In the previous chapters, regularisation was defined by various model constraints and applied additively to data fitting. This chapter presents another type of regularisation, which directly acts on the geophysical operator **G** so as to produce a stable solution estimate. This regularisation process is called the Backus-Gilbert method, named after its discoverers, geophysicists George E. Backus and James Freeman Gilbert (Backus and Gilbert, 1968, 1970).

For geophysical inverse problems, given a finite data set, there exist infinite solutions. Highly non-uniqueness of the inverse problems forces geophysicists to study generalised solutions by any means, instead of studying an exact solution. For instance, in the case of linear inversion we deal with

$$\int_{\Omega} G_k(r)\, x(r)\, \mathrm{d}r = d_k\,, \qquad (7.1)$$

where r is the position variable in the model space, k is the data sampling index, $k = 1, 2, \cdots, M$, and hence, $G_k(r)$ is an geophysical operator which, when acting on the model $x(r)$, will generate the data sample d_k. The geophysical field data are finite, $M < \infty$. The problem is ill-posed, and does not have a unique or exact solution. We can only find a solution estimate to fit the problem under some meaningful criteria.

The Backus-Gilbert method is a regularisation process for obtaining meaningful solutions to ill-posed inverse problems. Since possible solution estimates can be infinite, we should not be satisfied by finding a particular solution estimate. Instead, we should try to find many estimates and then

Seismic Inversion: Theory and Applications, First Edition. Yanghua Wang.
© 2017 John Wiley & Sons, Ltd. Published 2017 by John Wiley & Sons, Ltd.

appraise them in order to pick up the best from them as the 'final solution'. The Backus-Gilbert method is such an approach to analysing geophysical inverse problems with a finite and erroneous set of measurements, and to finding a unique average solution. Therefore, it is also known as the optimally localised average method.

7.1 The average solution

An average solution can be produced, as in the following,

$$\bar{x}(r_0) = \int_\Omega A(r, r_0)\, x(r)\, dr \,, \tag{7.2}$$

where $\bar{x}(r_0)$ is the average of the solution estimates at r_0, and $A(r, r_0)$ is an averaging function, called the averaging kernel, which should be unimodular,

$$\int_\Omega A(r, r_0)\, dr = 1 \,. \tag{7.3}$$

The Backus-Gilbert method suggests construction of the averaging kernel $A(r, r_0)$ as a linear combination of the known data kernels $\{G_k(r)\}$:

$$A(r, r_0) = \sum_{k=1}^{M} \alpha_k(r_0) G_k(r) \,, \tag{7.4}$$

where α_k are unknown weighting constants. Inserting Equation 7.4 into the integral in Equation 7.2 yields

$$\bar{x}(r_0) = \sum_{k=1}^{M} \alpha_k(r_0)(G_k(r), x(r)) = \sum_{k=1}^{M} \alpha_k(r_0)\, d_k \,. \tag{7.5}$$

This expression suggests that the solution $\bar{x}(r_0)$ is an average of data measurements $\{d_k\}$.

As the data set $\{d_k\}$ is incomplete, finding a unique solution for the Earth model from the inverse problem (Equation 7.1) is impossible. However, the average solution in Equation 7.5 can be uniquely determined from finite measurements $\{d_k\}$, as long as the unimodular averaging kernel A is a linear combination of the data kernels G_k.

7.2 The deltaness

The averaging kernel directly affects the resolution of a generalised solution. Hence, the averaging kernel $A(r,r_0)$ is a resolution function. When $A(r,r_0) \to \delta(r-r_0)$, i.e. the Dirac delta function, then $\bar{x}(r_0) \to x(r_0)$.

The averaging kernel $A(r,r_0)$ has particularly heavy weighting close to position r_0. *Deltaness* is a quantitative measure of the deviation of any arbitrary unimodular function $A(r,r_0)$ from an idealised $\delta(r-r_0)$ function.

To minimise the difference between the averaging kernel and the delta function, we may define an objective function in the form of an L_2 norm as

$$\phi_1(A,r_0) = \int_\Omega (A(r,r_0) - \delta(r-r_0))^2 dr . \tag{7.6}$$

Following definition in Equation 7.4, we have

$$\phi_1(\{\alpha_k\}, r_0) = \int_\Omega \left(\sum_{k=1}^M \alpha_k(r_0) G_k(r) - \delta(r-r_0) \right)^2 dr , \tag{7.7}$$

where $\{\alpha_k\}$ represents a series of α_k, for $k = 1, 2, \cdots, M$. Minimising this objective function by

$$\frac{\partial \phi_1}{\partial \alpha_k} = 2 \int_\Omega G_k(r) \left(\sum_{i=1}^M \alpha_i(r_0) G_i(r) - \delta(r-r_0) \right) dr = 0 , \tag{7.8}$$

we obtain a system of linear equations:

$$\sum_{i=1}^M \alpha_i(r_0) \int_\Omega G_i(r) G_k(r) dr = G_k(r_0) , \tag{7.9}$$

for $k = 1, 2, \cdots, M$. We should solve $\{\alpha_i(r_0)\}$, and then insert them into Equation 7.5 to produce a solution estimate $\bar{x}(r_0)$. This estimate is the quasi-solution.

7.3 The spread criterion

In Equation 7.6, let us denote $R(r) = A(r,r_0) - \delta(r-r_0)$. Equation 7.6 becomes the variance of $R(r)$ with respect to the mean zero. A small variance indicates that $R(r)$ values tend to be very close to the mean and hence to each other, whereas a high variance indicates that the $R(r)$ values are very different from the mean and from each other. However, this statistic measurement is the information about sample values, and does not reflect the spatial distribution of samples.

Figure 7.1 shows two spatial series that have the same variance, but one of which is better localised. We can measure localisation quantitatively by the width or by the spread.

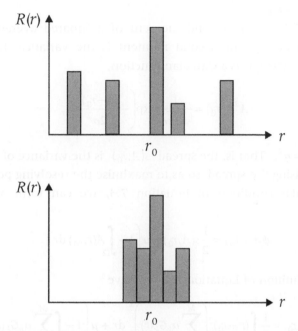

Figure 7.1 Two spatial series, $R(r)$, where r is the space coordinate, have the same variance, but the bottom case is better localised at r_0. Localisation can be measured quantitatively by the spread length, with respect to r_0.

The width or spread of $R(r)$ over the spatial coordinate r can be measured by a weighted variance:

$$s(R) = \int_{\Omega} w(r)R^2(r)\,\mathrm{d}r\,. \tag{7.10}$$

Backus and Gilbert (1968) suggest to define the weight by squared-distance, $w(r) = (r - r_0)^2$. The spread quantity is then expressed as

$$s(A,r_0) = \int_\Omega (r - r_0)^2 (A(r,r_0) - \delta(r - r_0))^2 \, dr \, . \tag{7.11}$$

Since $w(r_0)$ is zero, for $r = r_0$, we have

$$s(A,r_0) = \int_\Omega (r - r_0)^2 A^2(r,r_0) \, dr \, . \tag{7.12}$$

Therefore, this measurement is the 'spread of A from r_0', when $A(r,r_0)$ resembles $\delta(r - r_0)$.

Equation 7.12 is the second moment of a squared averaging kernel, $A^2(r,r_0)$. Statistically, the second moment is the variance. It is easy to verify that, if $A^2(r,r_0)$ is a Gaussian function,

$$A^2(r,r_0) = \frac{1}{\sqrt{2\pi\sigma}} \exp\left(-\frac{(r - r_0)^2}{2\sigma^2}\right), \tag{7.13}$$

then $s(A,r_0) = \sigma^2$. That is, the spread $s(A,r_0)$ is the variance of $A^2(r,r_0)$.

For minimising the spread, so as to maximise the resolving power, under the unimodular condition of Equation 7.3, we can form an objective function as

$$\phi_2(A,r_0) = \frac{1}{2} s(A,r_0) + \mu\left(1 - \int_\Omega A(r,r_0) \, dr\right). \tag{7.14}$$

Following definition of Equation 7.4, we have

$$\begin{aligned}
\phi_2(\{\alpha_k\},r_0) &= \frac{1}{2}\int_\Omega (r - r_0)^2 \left(\sum_{k=1}^{M} \alpha_k G_k(r)\right)^2 dr + \mu\left(1 - \int_\Omega \sum_{k=1}^{M} \alpha_k G_k(r) \, dr\right) \\
&= \frac{1}{2}\sum_{k=1}^{M}\sum_{i=1}^{M} \alpha_k(r_0)\alpha_i(r_0) H_{ki}(r_0) + \mu\left(1 - \sum_{k=1}^{M} \alpha_k(r_0) u_k\right),
\end{aligned} \tag{7.15}$$

where

$$H_{ki} = H_{ik} = \int_\Omega (r - r_0)^2 G_k(r) G_i(r) \, dr \, , \tag{7.16}$$

and

$$u_k = \int_\Omega G_k(r) \, dr \, . \tag{7.17}$$

The spread $s(A, r_0)$ can be written in a matrix-vector form as $s = \boldsymbol{\alpha}^T \mathbf{H} \boldsymbol{\alpha}$, and the unimodular constraint can be written as $1 - \boldsymbol{\alpha}^T \mathbf{u}$. Then, the objective function of Equation 7.15 can be written as

$$\phi_2(\boldsymbol{\alpha}, r_0) = \frac{1}{2} \boldsymbol{\alpha}^T \mathbf{H} \boldsymbol{\alpha} + \mu(1 - \boldsymbol{\alpha}^T \mathbf{u}) . \tag{7.18}$$

Minimising this objective function by setting $\partial \phi_2 / \partial \boldsymbol{\alpha} = \mathbf{0}$ yields a linear system of equations as

$$\mathbf{H} \boldsymbol{\alpha} = \mu \mathbf{u} . \tag{7.19}$$

Substituting

$$\boldsymbol{\alpha} = \mu \mathbf{H}^{-1} \mathbf{u} \tag{7.20}$$

into the unimodular constraint $\mathbf{u}^T \boldsymbol{\alpha} = 1$, we obtain

$$\mu = \frac{1}{\mathbf{u}^T \mathbf{H}^{-1} \mathbf{u}} . \tag{7.21}$$

Therefore, we can compute the weight constants by

$$\boldsymbol{\alpha}(r_0) = \frac{\mathbf{H}^{-1} \mathbf{u}}{\mathbf{u}^T \mathbf{H}^{-1} \mathbf{u}} . \tag{7.22}$$

Once we have calculated $\boldsymbol{\alpha}(r_0)$, we can use Equation 7.5 to estimate the averaged solution $\bar{x}(r_0)$ directly, $\bar{x}(r_0) = (\boldsymbol{\alpha}(r_0), \mathbf{d})$, where $\mathbf{d}$ is the data vector, and $(\boldsymbol{\alpha}, \mathbf{d}) = \boldsymbol{\alpha}^T \mathbf{d}$ is the inner product of two vectors.

7.4 The Backus-Gilbert stable solution

In the previous sections, we considered the finite measurements for the linear inverse problem in Equation 7.1, in which we also supposed that $\mathbf{d}$ have been measured without observational errors. For practical geophysical inversion we should consider that the geophysical field data are both finite and erroneous:

$$\tilde{\mathbf{d}} = \mathbf{d} + \mathbf{e} . \tag{7.23}$$

In this case, the solution estimate becomes

$$\tilde{x}(r_0) = (\boldsymbol{\alpha}(r_0), \tilde{\mathbf{d}}) = \bar{x}(r_0) + \delta x(r_0) , \tag{7.24}$$

where

$$\delta x(r_0) = (\boldsymbol{\alpha}(r_0), \mathbf{e}) . \tag{7.25}$$

Of course, we do not know the errors $\mathbf{e}$ (if we did, they would not be errors; we would remove them), but as in any error analysis we may assume that by repeated measurements we have learned something about the statistics of the errors. Now, we assume that the errors in the data are Gaussian, that is, independent of each other, and the mean is zero. We also assume that the variance matrix $\mathbf{E}$ of the errors of measurement can be estimated, and is symmetric and positive semidefinite.

The expectation of $\tilde{x}(r_0)$ is $\bar{x}(r_0)$. Then, the variance of the error in our solution estimate is

$$\text{var}[\tilde{x}(r_0)] = \text{var}[\delta x(r_0)] = \text{var}[(\boldsymbol{\alpha}(r_0), \mathbf{e})] = \boldsymbol{\alpha}^{\mathsf{T}}\mathbf{E}\boldsymbol{\alpha} , \tag{7.26}$$

where $\mathbf{E} = \mathbf{e}\mathbf{e}^{\mathsf{T}} = \tilde{\mathbf{d}}\tilde{\mathbf{d}}^{\mathsf{T}}$ is the variance matrix. This variance $\text{var}[\tilde{x}(r_0)]$ is not the expected deviation of $\tilde{x}(r_0)$ from the true solution $x(r_0)$, but rather measures the expected experiment-to-experiment scatter among estimates $\tilde{x}(r_0)$ if the whole experiment were to be repeated many times.

Since we know the data variance matrix $\mathbf{E}$, the variance of the solution estimates is completely determined by the coefficients $\boldsymbol{\alpha}$. Therefore, we could choose a positive number ε and refuse to consider any averaging kernels unless they satisfied

$$\boldsymbol{\alpha}^{\mathsf{T}}\mathbf{E}\boldsymbol{\alpha} < \varepsilon^2 . \tag{7.27}$$

We should minimise this variance of solution estimates.

Combining the variance and the spread, together with the unimodular constraint, we have an objective function as

$$\phi_3(\boldsymbol{\alpha}) = \frac{1}{2}(\boldsymbol{\alpha}^{\mathsf{T}}\mathbf{H}\boldsymbol{\alpha} + \lambda\boldsymbol{\alpha}^{\mathsf{T}}\hat{\mathbf{E}}\boldsymbol{\alpha}) + \mu(1 - \boldsymbol{\alpha}^{\mathsf{T}}\mathbf{u}) , \tag{7.28}$$

where $\hat{\mathbf{E}} = c\mathbf{E}$ is a scaled variance matrix $\mathbf{E}$ with a positive constant c, so that $\mathbf{H}$ and $\hat{\mathbf{E}}$ will be of comparable numerical size, and λ is a trade-off parameter between $\mathbf{H}$ and $\hat{\mathbf{E}}$.

Unfortunately, the minimisation of the spread is contradictory to the minimisation of the solution variance. Therefore, we should try a best trade-off between them. Backus and Gilbert (1970) suggested replacing λ with $\tan\theta$ and rewriting the objective function as

$$\phi_3(\boldsymbol{\alpha}) = \frac{1}{2}(\boldsymbol{\alpha}^T \mathbf{H} \boldsymbol{\alpha} \cos\theta + \boldsymbol{\alpha}^T \hat{\mathbf{E}} \boldsymbol{\alpha} \sin\theta) + \mu(1 - \boldsymbol{\alpha}^T \mathbf{u}), \qquad (7.29)$$

where $\theta \in [0, \frac{1}{2}\pi]$ is a new trade-off parameter.

We shall first find the unknown weights $\boldsymbol{\alpha}$, by minimising $\phi_3(\boldsymbol{\alpha})$. Setting $\partial\phi_3 / \partial\boldsymbol{\alpha} = \mathbf{0}$, we have

$$\boldsymbol{\alpha} = \mu(\mathbf{H} \cos\theta + \hat{\mathbf{E}} \sin\theta)^{-1} \mathbf{u}. \qquad (7.30)$$

Because of the unimodular constraint $\mathbf{u}^T \boldsymbol{\alpha} = 1$, we get

$$\mu = \frac{1}{\mathbf{u}^T (\mathbf{H} \cos\theta + \hat{\mathbf{E}} \sin\theta)^{-1} \mathbf{u}}. \qquad (7.31)$$

Finally, we obtain the weights,

$$\boldsymbol{\alpha} = \frac{(\mathbf{H} \cos\theta + \hat{\mathbf{E}} \sin\theta)^{-1} \mathbf{u}}{\mathbf{u}^T (\mathbf{H} \cos\theta + \hat{\mathbf{E}} \sin\theta)^{-1} \mathbf{u}}. \qquad (7.32)$$

Equations 7.31 and 7.32 for μ and $\boldsymbol{\alpha}$, when $\theta \neq 0$, are extensions of Equations 7.21 and 7.22, respectively.

Then, we shall find the best trade-off parameter θ. Given different values for the trade-off parameter $\theta \in [0, \frac{1}{2}\pi]$, we can use Equation 7.32 to calculate the coefficients $\boldsymbol{\alpha}(\theta)$.

For each of coefficients $\boldsymbol{\alpha}(\theta)$, we can make a solution estimate $\tilde{x}(r_0, \theta) = (\boldsymbol{\alpha}(r_0, \theta), \tilde{\mathbf{d}})$, according to Equation 7.5. Theoretically, all of these solution estimates are acceptable to some extent. However, the existence of so many solution estimates reflects the non-uniqueness of the practical inverse problems. It is important to extract the principle property of these solution estimates, $\tilde{x}(r_0, \theta)$, the only information that can be obtained from the finite erroneous data.

With different values $\boldsymbol{\alpha}(\theta)$ for the trade-off parameter $\theta \in [0, \frac{1}{2}\pi]$, we have the spread as

$$s(\theta) = \boldsymbol{\alpha}^T(\theta) \mathbf{H} \boldsymbol{\alpha}(\theta), \qquad (7.33)$$

and the variance as

$$\varepsilon^2(\theta) = \boldsymbol{\alpha}^T(\theta) \mathbf{E} \boldsymbol{\alpha}(\theta). \qquad (7.34)$$

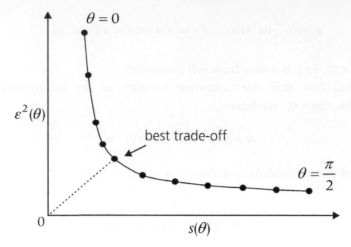

Figure 7.2   A trade-off curve for appraising solution estimates. The horizontal axis is the spread (or the resolution width) and the vertical axis is the variance of the solution estimates. The best trade-off occurs at the most closest point to the origin in the curve whose corresponding value of $\theta = \theta_0$ is the best trade-off parameter.

The relationship between the resolution width $s(\theta)$ and the variance of the solution estimate $\varepsilon^2(\theta)$ is plotted in Figure 7.2.

- When $\theta = 0$, we minimise the spread;

- When $\theta = \frac{1}{2}\pi$, we minimise the solution variance;

- When at the origin, zero resolution width means the highest resolution, and zero variance represents the best solution estimate;

- The best trade-off occurs at the most closest point to the origin in the curve whose corresponding value of $\theta = \theta_0$ is the best trade-off parameter.

Once we select the best trade-off parameter θ_0, we can estimate $\boldsymbol{\alpha}(r_0, \theta_0)$, and finally the average solution at position r_0 by

$$\tilde{x}(r_0) = (\boldsymbol{\alpha}(r_0, \theta_0), \tilde{\mathbf{d}}) . \tag{7.35}$$

In summary, the Backus-Gilbert method imposes a stability constraint to the objective function. It seeks to maximise the stability of the solution, so that the solution would vary as little as possible if the input data were resampled multiple times. This is in contrast to other regularisation methods, such as Tikhonov regularisation, which seeks to impose smoothness constraints on the solution. However, a *stable* solution is almost inevitably bound to be *smooth*. In practice, the stability constraint in the Backus-Gilbert method will result in a smooth solution, the so-called averaged solution.

CHAPTER 8

Seismic wavelet estimation

For a profound understanding on the fundamental theory and concepts, with emphasis on the physical meanings of linearised inverse problems, we shall demonstrate their practical applications in seismic inversion in the following chapters. In this book, we shall categorise these applications systematically into two groups, based on either the Earth convolutional model or the wave equation model.

Both inversion categories need to know seismic wavelet, which combines source signature, receiver response, recording instrument effect, and importantly wave distortion within the subsurface media, and so on. In the convolutional model, a seismic trace $d(t)$ is expressed as

$$d(t) = w(t) * r(t) , \qquad (8.1)$$

where $w(t)$ is the seismic wavelet, and $r(t)$ is the reflectivity series. If we fortunately have well-log information available, we should be able to estimate a wavelet by correlation between a seismic trace at the vicinity of the well location and a reflectivity series calculated, based on well-logs.

If there is no well-log available, it is an inverse problem to estimate essentially two unknowns, the wavelet and the reflectivity series, from this single equation. Practically, we need to make a reliable wavelet estimate first, independently from reflectivity inversion, before attempting to retrieve the reflectivity series from seismic traces.

In order to estimate the wavelet from seismic traces, we need to assume that the reflectivity series has a flat power spectrum, so that we can use the power spectrum of seismic data to approximately represent the power spectrum of the wavelet. In the current chapter, we will introduce a

Seismic Inversion: Theory and Applications, First Edition. Yanghua Wang.
© 2017 John Wiley & Sons, Ltd. Published 2017 by John Wiley & Sons, Ltd.

wavelet construction method based on the statistical characteristics of the power spectrum, and two inversion methods for a constant-phase wavelet and a mixed phase wavelet, respectively.

8.1 Wavelet extraction from seismic-to-well correlation

While a field seismic trace has noise, computed reflectivity series will not be error-free either:

$$\tilde{d}(t) = d(t) + e(t) \,,$$

$$\tilde{r}(t) = r(t) + \varepsilon(t) \,,$$
(8.2)

where $e(t)$ is noise in the seismic observation, and $\varepsilon(t)$ is the error in the estimate of the reflectivity series. The cross-covariance sequence between $\tilde{r}(t)$ and $\tilde{d}(t)$ can be estimated by

$$\tilde{\phi}_{\mathrm{rd}}(\tau) = \mathrm{cov}\{\tilde{r}(t), \tilde{d}(t+\tau)\} = \frac{1}{M} \sum_t \tilde{r}_t \tilde{d}_{t+\tau} \,,$$
(8.3)

where M is the total number of time samples, and τ is the time shift. The auto-covariance sequences for $\tilde{d}(t)$ and $\tilde{r}(t)$, respectively, are

$$\tilde{\phi}_{\mathrm{dd}}(\tau) = \mathrm{cov}\{\tilde{d}(t), \tilde{d}(t+\tau)\} = \frac{1}{M} \sum_t \tilde{d}_t \tilde{d}_{t+\tau} \,,$$
(8.4)

$$\tilde{\phi}_{\mathrm{rr}}(\tau) = \mathrm{cov}\{\tilde{r}(t), \tilde{r}(t+\tau)\} = \frac{1}{M} \sum_t \tilde{r}_t \tilde{r}_{t+\tau} \,.$$
(8.5)

Fourier transforming $\tilde{\phi}_{\mathrm{rd}}(\tau)$, $\tilde{\phi}_{\mathrm{dd}}(\tau)$ and $\tilde{\phi}_{\mathrm{rr}}(\tau)$ generates the cross- and auto-spectra, $\tilde{\Phi}_{\mathrm{rd}}(\omega)$, $\tilde{\Phi}_{\mathrm{rr}}(\omega)$ and $\tilde{\Phi}_{\mathrm{dd}}(\omega)$, respectively.

For a noise-free model $d(t) = w(t) * r(t)$, these cross- and auto-spectra $\Phi_{\mathrm{rd}}(\omega)$, $\Phi_{\mathrm{dd}}(\omega)$, and $\Phi_{\mathrm{rr}}(\omega)$ follow the relationships,

$$\Phi_{\mathrm{rd}}(\omega) = W(\omega)\Phi_{\mathrm{rr}}(\omega) \,,$$

$$\Phi_{\mathrm{dd}}(\omega) = |W(\omega)|^2 \, \Phi_{\mathrm{rr}}(\omega) \,,$$
(8.6)

$$|\Phi_{\mathrm{rd}}(\omega)|^2 = \Phi_{\mathrm{dd}}(\omega)\Phi_{\mathrm{rr}}(\omega) \,,$$

where $W(\omega)$ is the frequency spectrum of the wavelet $w(t)$.

The relationships of the spectra between the noisy and noise-free data are

$$\tilde{\Phi}_{rd}(\omega) = \Phi_{rd}(\omega) \,,$$

$$\tilde{\Phi}_{dd}(\omega) = \Phi_{dd}(\omega) + \Phi_{ee}(\omega) \,, \tag{8.7}$$

$$\tilde{\Phi}_{rr}(\omega) = \Phi_{rr}(\omega) + \Phi_{\varepsilon\varepsilon}(\omega) \,.$$

Based on the relations in Equations 8.6 and 8.7, we can derive a formula for estimating wavelet $W(\omega)$ in the frequency domain. The ratio of cross- and auto-spectra,

$$\frac{\tilde{\Phi}_{rd}}{\tilde{\Phi}_{rr}} = \frac{W\,\Phi_{rr}}{\Phi_{rr} + \Phi_{\varepsilon\varepsilon}} \,, \tag{8.8}$$

leads to

$$W = \left(\frac{\Phi_{rr}}{\Phi_{rr} + \Phi_{\varepsilon\varepsilon}}\right)^{-1} \frac{\tilde{\Phi}_{rd}}{\tilde{\Phi}_{rr}} \,. \tag{8.9}$$

The magnitude-squared coherence, which measures the linear correlation between $\tilde{r}(t)$ and $\tilde{d}(t)$, is defined as

$$\rho^2 = \frac{|\tilde{\Phi}_{rd}|^2}{\tilde{\Phi}_{rr}\tilde{\Phi}_{dd}} \,. \tag{8.10}$$

It can be expanded to

$$\rho^2 = \frac{\Phi_{rr}}{\Phi_{rr} + \Phi_{\varepsilon\varepsilon}} \frac{\Phi_{dd}}{\Phi_{dd} + \Phi_{ee}} \,. \tag{8.11}$$

Comparing Equations 8.10 and 8.11, we obtain

$$\left(\frac{\Phi_{rr}}{\Phi_{rr} + \Phi_{\varepsilon\varepsilon}}\right)^{-1} = \frac{1}{1+\mu} \frac{\tilde{\Phi}_{rr}\tilde{\Phi}_{dd}}{|\tilde{\Phi}_{rd}|^2} \,, \tag{8.12}$$

where $\mu = \Phi_{ee}/\Phi_{dd}$ is the noise-to-signal ratio of the seismic trace, and can be estimated by the multichannel coherence analysis to the replicated seismic traces (White, 1984). Then, we have the following formula for the wavelet estimation:

$$W = \frac{1}{1+\mu} \frac{\tilde{\Phi}_{dd}}{\tilde{\Phi}_{rd}} \,. \tag{8.13}$$

The time-domain wavelet $w(t)$ is given by the inverse Fourier transform of $W(\omega)$. It should be tapered by a window function. Here, we propose a window function as the following:

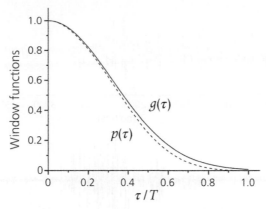

Figure 8.1 Two window functions. A Gaussian function $g(\tau)$ (the solid curve) with the standard deviation $\sigma = T/\pi$, where T is a half of the expected wavelet length, and the Parzen window function $p(\tau)$ (the dashed curve).

$$g(\tau) = \exp\left(-\frac{1}{2}\left(\frac{\pi\tau}{T}\right)^2\right), \tag{8.14}$$

where T is a half of the expected wavelet length. This continuous window function (the solid curve in Figure 8.1) is in the Gaussian form with the standard deviation of $\sigma = T/\pi$, and is close to the Parzen window function (the dashed curve in Figure 8.1), defined as

$$p(\tau) = \begin{cases} 1 - 6\left(\dfrac{\tau}{T}\right)^2\left(1 - \dfrac{|\tau|}{T}\right), & |\tau| \le \dfrac{1}{2}T, \\[2ex] 2\left(1 - \dfrac{|\tau|}{T}\right)^3, & \dfrac{1}{2}T < |\tau| \le T, \\[2ex] 0, & |\tau| > T. \end{cases} \tag{8.15}$$

Figure 8.2 displays a seismic trace at a well location, a reflectivity series calculated based on well-log information, and the wavelet extracted from well-seismic correlation, using Equation 8.13.

In practice, the drill location of the well with respect to the seismic volume may not be optimal for extracting the wavelet. We need to search for a seismic-well tie location within a specified vicinity of the posted well location (White *et al.*, 1998; White and Simm, 2003). The calibration processing consists of the following steps:

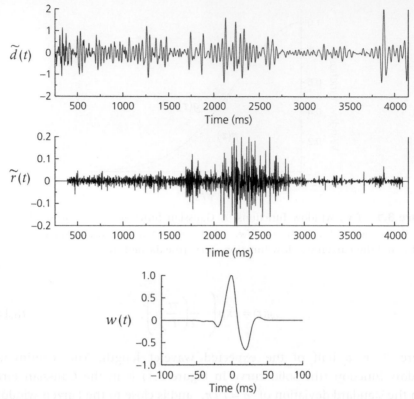

Figure 8.2 A seismic trace (top) at a well location, a reflectivity series (the middle) calculated based on well-log information, and the wavelet (the bottom) extracted from well-seismic correlation.

1) Applying a band-pass filter to the reflectivity series, $\hat{r}(t) = b(t) * \tilde{r}(t)$, where $b(t)$ is a filter, and $\tilde{r}(t)$ is the broadband reflectivity series based on the product of sonic and density log data. The filtered time series $\hat{r}(t)$ should have the bandwidth being roughly same as the seismic trace.

2) Applying the inverse Fourier transform to the spectral coherence function, so as to find the best time shift associated with each trace. The best spatial location is selected among traces in the vicinity of the posted well.

Then, an optimal wavelet can be estimated using the method above, based on the trace $\tilde{d}(t)$ from the best spatial location and aligned reflectivity series $\tilde{r}(t)$ with the best lag.

8.2 Generalised wavelet constructed from power spectrum

When there is no well-log information, for the wavelet estimation, it often assumes that the reflectivity series is a sequence of serially uncorrelated random variables with zero mean and finite variance. The random variables have equal powers within a fixed bandwidth at any centre frequency. In other words, the reflectivity series is often assumed to be white or nearly so, and has a flat power spectral density.

Then, the power spectrum of a wavelet approximately equals to the power spectrum of the seismic trace:

$$| W(\omega) |^2 = \frac{1}{\gamma^2} | D(\omega) |^2, \qquad (8.16)$$

where $\gamma^2 = | R(\omega) |^2$ is the power spectrum of the reflectivity series.

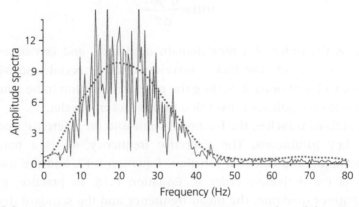

Figure 8.3 The amplitude spectrum of a seismic trace (the solid curve) and the amplitude spectrum of a wavelet (the dotted curve). Both the seismic trace and the estimated wavelet were shown in Figure 8.2.

Figure 8.3 demonstrates that the amplitude spectrum of a seismic trace $| D(\omega) |$ and the spectrum of an estimated wavelet $| W(\omega) |$ are close to each other. Both the seismic trace and the wavelet are shown in Figure 8.2.

Whereas we treat the power spectrum of a seismic trace as the spectrum of its wavelet, we can extract the wavelet robustly from the power spectrum of field seismic data, using a concept of generalised wavelets.

The generalised seismic wavelets are presented in a concise form analytically in the time domain. The well-known Ricker wavelet is a symmetrical waveform in the time domain (Ricker, 1953). However, practically observed seismic signals are in non-Ricker forms. Wang (2015b) suggest to systematically generalise the symmetric Ricker wavelet to be asymmetrical, in order to better represent practical seismic signals.

Generalised wavelets are defined mathematically as fractional derivatives of a Gaussian function. We set the potential function in a (negative) Gaussian:

$$g(t) = -\sqrt{\pi}\,\omega_0 \exp\left(-\frac{\omega_0^2}{4}(t - \tau_0)^2\right), \qquad (8.17)$$

where τ_0 is the time position of the symmetrical centre, and ω_0 is a reference frequency (in radians per second). The wavelets are defined as the fractional derivative of this potential function:

$$w(t) \equiv \frac{\mathrm{d}^u g(t)}{\mathrm{d}t^u}, \qquad (8.18)$$

where u is the order of a time-domain derivative, and can be either a fraction or an integer. The Ricker wavelet is just a special case with the integer derivative of order 2. Setting the potential function to be a negative Gaussian is for the polarity convention of this Ricker wavelet.

In generalised wavelets, the fractional value and the reference frequency are two key parameters. The reference frequency ω_0 is a parameter needed to define the potential function, Equation 8.17, and the fractional value u is the derivative order in Equation 8.18. In practice, given a discrete Fourier spectrum, the mean frequency and the standard deviation can be evaluated numerically. Once these two quantities are measured statistically from field seismic data, the fractional value and the reference frequency can be uniquely determined, and in turn the wavelet is analytically derived.

Considering a 2-D seismic profile (Figure 8.4a) as a single time series, by linking all traces to each other, the reflectivity series can be approximated as a series of random numbers. Thus, the frequency spectrum $A(\omega)$ (the fluctuated curve in Figure 8.4b) shall reflect the property of the wavelet.

For any real-valued seismic signal, the frequency spectrum $A(\omega)$ needs to be considered only for $\omega \geq 0$. Based on the power spectrum $|A(\omega)|^2$, the mean frequency and the standard deviation may be evaluated by using (Berkhout, 1984; Cohen and Lee, 1989)

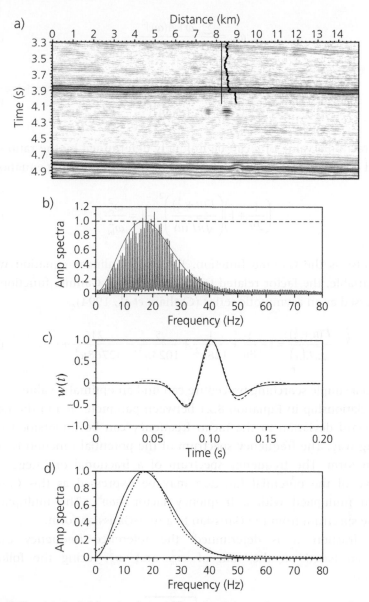

Figure 8.4 (*a*) A field seismic profile, consisting of 600 traces. The vertical straight line indicates a vertical borehole, and the black curve is the acoustic impedance calculated from logging information. (*b*) The amplitude spectrum of a seismic profile (the fluctuated curve) and the spectrum of the generalised wavelet (the smooth curve). (*c*) The wavelet extracted using the generalised wavelet method (the solid curve) and a constant-phase wavelet (the dashed curve) extracted from the well-seismic correlation. (*d*) The spectra of the generalised wavelet (the solid curve) and the constant-phase wavelet (the dashed curve).

$$\omega_m = \frac{\sum_\omega \omega \mid A(\omega) \mid^2}{\sum_\omega \mid A(\omega) \mid^2} , \qquad (8.19)$$

$$\omega_\sigma = \left(\frac{\sum_\omega (\omega - \omega_m)^2 \mid A(\omega) \mid^2}{\sum_\omega \mid A(\omega) \mid^2} \right)^{1/2} . \qquad (8.20)$$

The fractional value of u can be uniquely determined by the ratio of the standard deviation to the mean frequency, using the following equation:

$$\left(\frac{1}{2u} + 1 \right) \left(\frac{\Gamma(u + \frac{1}{2})}{\sqrt{u}\Gamma(u)} \right)^2 - 1 = \frac{\omega_\sigma^2}{\omega_m^2} , \qquad (8.21)$$

where $\Gamma(s)$ is the Gamma function. For this nonlinear equation with a single variable, the factor related to the ratio of the Gamma functions can be expressed as an asymptotic series (Graham *et al.*, 1994),

$$\frac{\Gamma(u + \frac{1}{2})}{\sqrt{u}\Gamma(u)} = 1 - \frac{1}{8u} + \frac{1}{128u^2} + \frac{5}{1024u^3} - \frac{21}{32768u^4} + \cdots \qquad (8.22)$$

and then a simple searching procedure can find an optimal u value.

The relationship in Equation 8.21 between parameter u and the ratio of the standard deviation to the mean frequency can be understood in the following way. The frequency spectrum of the potential function is also in Gaussian form. The frequency spectrum of a fractional or integer order derivative of the potential function may be expressed as this Gaussian spectrum multiplied with a frequency factor $(i\omega)^u$. The multiplication alters the spectrum from the Gaussian to a non-Gaussian form.

Once fraction u is determined, the reference frequency can be determined based on the sum of ω_m^2 and ω_σ^2, using the following expression:

$$\omega_0 = 2\sqrt{\frac{\omega_m^2 + \omega_\sigma^2}{1 + 2u}} . \qquad (8.23)$$

This expression suggests to use both parameters ω_m and ω_σ, instead of only ω_m or only ω_σ, for estimating ω_0. It is an effort to minimise any potential bias error in these two values, which have been numerically evaluated from a discrete Fourier spectrum.

Once the fractional value u and the reference frequency ω_0 are obtained, the frequency spectrum of a generalised wavelet is set analytically as (Wang, 2015b)

$$W(\omega) = \left(\frac{u}{2}\right)^{-u/2} \frac{\omega^u}{\omega_0^u} \exp\left(-\frac{\omega^2}{\omega_0^2} + \frac{u}{2}\right) \exp\left(-i\omega\tau_0 + i\pi\left(1 + \frac{u}{2}\right)\right). \quad (8.24)$$

Performing an inverse Fourier transform on $W(\omega)$, we can generate a time-domain wavelet $w(t)$.

The generalised wavelet (the solid curve in Figure 8.4c) is close to the form of a constant-phase wavelet extracted from well-seismic tie (the dashed curve in Figure 8.4c). The adjusted constant phase is $\theta_0 = -18°$, estimated using the kurtosis matching method (in the next section). Note that the amplitude spectra between these two wavelets (Figure 8.4d) are close but not identical.

The reference frequency ω_0 is inversely proportional to the deviation of the Gaussian distribution, but is not the peak frequency, ω_p. The relationship between ω_0 and ω_p is

$$\omega_p = \omega_0 \sqrt{\frac{u}{2}}. \quad (8.25)$$

Only for the case of $u = 2$, the Ricker wavelet, the reference frequency ω_0 equals to the peak frequency ω_p (Wang, 2015a, c). In reality, seismic signals are often not symmetric (Hosken, 1988) and, instead, are close to the fractional derivatives of a Gaussian.

8.3 Kurtosis matching for a constant-phase wavelet

As demonstrated in the previous section, the phase is the all-important factor in the wavelet estimation. According to Equation 8.24, generalised wavelets have a constant phase (independent from the frequency). We can also estimate the phase explicitly through inversion.

For estimating the phase of the wavelet, certain statistical properties in seismic data need to be exploited. In this section, we introduce a kurtosis matching method, for estimating a constant phase of the wavelet, and in the next section we will introduce a cumulant matching method, for a mixed-phase wavelet.

If each sample of a series has a normal distribution with zero mean, this series is said to be Gaussian white. In fact, seismic reflectivity series has a leptokurtic amplitude distribution, which means the distribution of the sequence is more heavy-tailed than a Gaussian one. Convolving this white reflectivity series with an arbitrary wavelet renders the outcome seismic trace less white, but more Gaussian. In statistics, kurtosis can measure the deviation from Gaussianity. Maximising the kurtosis means to recover the leptokurtic feature of the original reflectivity series, and less formally, to recover the spikiness of the sequence.

For a zero-mean seismic trace, the kurtosis is defined as

$$K_{\mathrm{d}} = \frac{\mathrm{E}\{d_t^4\}}{(\mathrm{var}\{d_t\})^2} ,$$
(8.26)

where $\mathrm{E}\{\cdot\}$ is the expectation, and $\mathrm{var}\{d_t\} = \mathrm{E}\{d_t^2\}$ is the variance of d_t. Therefore, kurtosis is the standardised fourth moment of a probability distribution: the fourth moment about the mean divided by the square of its variance. Kurtosis K_{d} defined in Equation 8.26 is the population value. A sample value can be estimated from a seismic trace by

$$\hat{K}_{\mathrm{d}} = \frac{\sum d_t^4 / M}{(\sum d_t^2 / M)^2} ,$$
(8.27)

where $\hat{K}_{\mathrm{d}}$ with circumflex indicates calculated kurtosis being a sample value, and M is the total number of samples of d_t.

The kurtosis of a seismic reflectivity profile shown in Figure 8.5a is $\hat{K}_{\mathrm{d}} = 23.3$. The convolution of this reflectivity profile with a wavelet generates a synthetic seismic profile (Figure 8.5b). The kurtosis of the seismic profile is reduced drastically to $\hat{K}_{\mathrm{d}} = 5.9$. Therefore, the higher this quantity is, the more spiky the sequence $\{d_t\}$ will be.

Figure 8.6 displays constant-phase wavelets with different phase angles, and the corresponding kurtosis values of seismic profiles, generated with these wavelets. The seismic profile with a zero-phase wavelet has the largest kurtosis value. Therefore, if applying a phase rotation θ to any seismic profile, and calculating the corresponding $\hat{K}_{\mathrm{d}}(\theta)$, an optimum phase rotation $\hat{\theta}$ can be found. Corresponding to the maximum kurtosis, the phase-shifted profile will be spiky, and the phase of the rotated wavelet will be close to zero phase.

The seismic trace $d(t)$ is normalised to unit variance and zero mean, and is denoted by $x(t)$. A quadrature trace can be formed by Hilbert

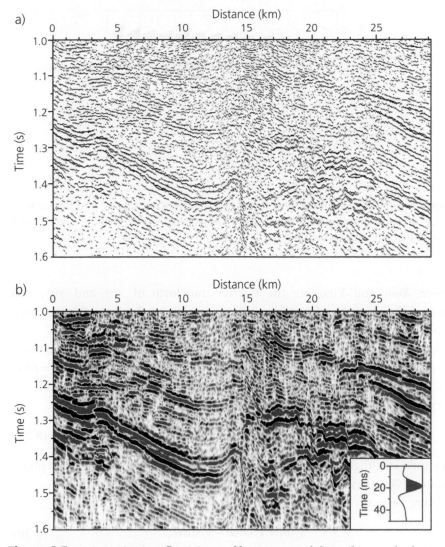

Figure 8.5 (*a*) A seismic reflectivity profile. Because of the spikiness, the kurtosis of this reflectivity profile is very high $(\hat{K}_d = 23.3)$. (*b*) A seismic section generated by convolution of the reflectivity profile and a wavelet (in the corner). After the convolution, the kurtosis is reduced drastically $(\hat{K}_d = 5.9)$.

transform, $y(t) = H[x(t)]$. In the frequency domain, the Hilbert transform is a multiplier:

$$Y(\omega) = -\text{i} \, \text{sgn}(\omega) \, X(\omega) \,, \tag{8.28}$$

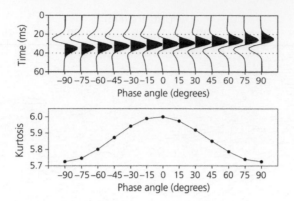

Figure 8.6 Constant-phase wavelets with different phase angles, and the kurtosis values of corresponding seismic profiles. The seismic profile with a zero-phased wavelet has the largest kurtosis value.

where $X(\omega)$ and $Y(\omega)$ are the Fourier transform of $x(t)$ and $y(t)$, and sgn is the signum function, $\mathrm{sgn}(\omega) = -1,\ 0,\ 1,$ for $\omega < 0,\ = 0,\ > 0,$ respectively.

As seismic trace $x(t)$ is generally real valued, its Fourier transform $X(-\omega) = \overline{X}(\omega)$ and $X(\omega = 0) = 0$. After Hilbert transform, $Y(-\omega) = \overline{Y}(\omega)$ and $Y(\omega = 0) = 0$, hence the time-domain trace $y(t)$ is also real valued.

For $\omega > 0$, the Hilbert operator is

$$- \mathrm{i}\, \mathrm{sgn}(\omega) = -\mathrm{i} = e^{-\mathrm{i}\pi/2}. \tag{8.29}$$

This indicates that the quadratic trace $y(t)$ is a version with a $-90°$ phase rotation ($90°$ phase lag) from trace $x(t)$.

Using $x_t = x(t)$ and $y_t = y(t)$, a composite trace can be built as

$$z_t(\theta) = x_t \cos\theta - y_t \sin\theta, \tag{8.30}$$

where θ is a phase shift relative to x_t. Since $\sum x_t^2 / M = 1$ and the Hilbert transform operator has a power of unity, then $\sum y_t^2 / M = 1$. At zero-lag, x_t and y_t are uncorrelated, $\sum z_t^2 / M = 1$. Therefore, the kurtosis of the composed trace is measured by

$$\hat{K}_z(\theta) = \frac{1}{M} \sum z_t^4(\theta). \tag{8.31}$$

The maximum kurtosis phase estimation method (White, 1988; Longbottom *et al.*, 1988) calculates a series of kurtosis values with respect to different phase angles, and locates a constant-phase rotation $\hat{\theta}$ that maximises the non-Gaussianity of the seismic trace.

If u_t is the trace corresponding to a zero-phase signal, and v_t is the Hilbert transform ($-90°$ phase shift) of u_t, the input trace x_t and the quadrature trace y_t can be represented as

$$x_t = u_t \cos\theta_0 - v_t \sin\theta_0 ,$$
$$y_t = u_t \sin\theta_0 + v_t \cos\theta_0 , \tag{8.32}$$

where θ_0 is the constant phase of the existing signal. Then the composite trace $z_t = x_t \cos\theta - y_t \sin\theta$ with phase shift θ is

$$z_t = u_t \cos(\theta + \theta_0) - v_t \sin(\theta + \theta_0) . \tag{8.33}$$

If $\theta + \theta_0 = 0$, then $z_t = u_t$. Therefore, the most likely wavelet phase θ_0 is corresponding to the negative value of angle $\hat{\theta}$ that maximises the kurtosis value. The phase of the wavelet can be approximated to be a constant

$$\theta_0 = -\hat{\theta} . \tag{8.34}$$

Figure 8.7 is the kurtosis variation with rotation angle applied to the synthetic seismic profile (shown in Figure 8.5b). The maximum kurtosis occurs at $\hat{\theta} = -30°$. The wavelet then has an estimated phase of $\theta_0 = 30°$.

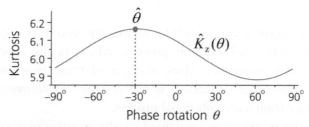

Figure 8.7 The kurtosis $\hat{K}_z(\theta)$ versus the scanning rotation angle θ, applied to the seismic profile. The maximum kurtosis occurs at $\hat{\theta}$, which is the negative value of the wavelet phase angle θ_0.

8.4 Cumulant matching for a mixed-phase wavelet

The cumulant is also a statistic concept. The kth-order cumulant function of a real stationary process is given by

$$c_k^x(\tau_1,\tau_2,\cdots,\tau_{k-1}) = m_k^x(\tau_1,\tau_2,\cdots,\tau_{k-1}) - m_k^g(\tau_1,\tau_2,\cdots,\tau_{k-1}), \qquad (8.35)$$

where m_k^x is the kth-order moment function of a zero-mean seismic trace x_t,

$$m_k^x(\tau_1,\tau_2,\cdots,\tau_{k-1}) = E\{x_t x_{t+\tau_1} \cdots x_{t+\tau_{k-1}}\}, \qquad (8.36)$$

and m_k^g is the moment function of an equivalent Gaussian process, g_t, that has the same second-order statistics as x_t:

$$m_k^g(\tau_1,\tau_2,\cdots,\tau_{k-1}) = E\{g_t g_{t+\tau_1} \cdots g_{t+\tau_{k-1}}\}. \qquad (8.37)$$

Therefore, the cumulant not only displays the amount of higher-order correlation (the first term on the right-hand side), but also provides a measure of the distance of this random process from Gaussianity.

Explicitly, the second-order cumulant of a zero-mean process is identical to the auto-correlation:

$$c_2^x(\tau) = \frac{1}{M}\sum_t x_t x_{t+\tau}, \qquad (8.38)$$

which is a phaseless function. The third-order cumulant is

$$c_3^x(\tau_1,\tau_2) = \frac{1}{M}\sum_t x_t x_{t+\tau_1} x_{t+\tau_2}. \qquad (8.39)$$

This two-dimensional function will be identically zero if its argument is a symmetrically distributed random process, which is the case for most seismic reflectivity sequences. Thus, the use of these two statistics (the second- and third-order cumulant) is not suitable for recovering a non-minimum phase from a convolutional process.

However, the information contained in the fourth-order cumulant is enough for determining the wavelet within a polarity reversal and a time shift when the reflectivity process is non-Gaussian. The fourth-order cumulant function is

$$c_4^x(\tau_1,\tau_2,\tau_3) = \frac{1}{M}\sum_t x_t x_{t+\tau_1} x_{t+\tau_2} x_{t+\tau_3} - c_2^x(\tau_1)c_2^x(\tau_2 - \tau_3)$$
$$\qquad (8.40)$$
$$-c_2^x(\tau_2)c_2^x(\tau_3 - \tau_1) - c_2^x(\tau_3)c_2^x(\tau_1 - \tau_2).$$

Considering the convolutional model, $x_t = w_t * r_t + e_t$, where e_t is an additive error series, the relationship among the fourth-order statistics for the convolutional model (Mendel, 1991) is

$$c_4^x(\tau_1, \tau_2, \tau_3) = c_4^r(\tau_1, \tau_2, \tau_3) * m_4^w(\tau_1, \tau_2, \tau_3) + c_4^e(\tau_1, \tau_2, \tau_3), \qquad (8.41)$$

where $c_4^r(\tau_1, \tau_2, \tau_3)$ and $c_4^e(\tau_1, \tau_2, \tau_3)$ are the fourth-order cumulant of the reflectivity sequence and the error series, respectively, and $m_4^w(\tau_1, \tau_2, \tau_3)$ is the fourth-order moment of the wavelet. If we assume that r_t is independent, identically distributed and non-Gaussian, and that the error trace e_t is in Gaussian (but does not need to be white), it can be rewritten as

$$c_4^x(\tau_1, \tau_2, \tau_3) = \gamma_4^r m_4^w(\tau_1, \tau_2, \tau_3), \qquad (8.42)$$

where $\gamma_4^r = c_4^x(0, 0, 0)$ is the kurtosis of the reflectivity series. That is, the fourth-order moment of the wavelet equals to, within a scale factor, the fourth-order cumulant of the trace.

In practice, neither c_4^e nor c_4^r is a multidimensional spike at zero lag. The cumulant estimate can be improved by applying a three-dimensional (3-D) smooth-tapering window to the seismic trace cumulant, especially for those cases in which c_4^x is very poor because of the low amount of data available. Then, the fourth-order cumulant estimate of the wavelet $\hat{m}_4^w$ is given by

$$\hat{m}_4^w(\tau_1, \tau_2, \tau_3) = \frac{1}{\gamma_4^r} a(\tau_1, \tau_2, \tau_3) c_4^x(\tau_1, \tau_2, \tau_3), \qquad (8.43)$$

where $a(\tau_1, \tau_2, \tau_3)$ is a smoothing window function. A 3-D window function can be written as

$$a(\tau_1, \tau_2, \tau_3) = g(\tau_1) g(\tau_2) g(\tau_3) g(\tau_2 - \tau_1) g(\tau_3 - \tau_2) g(\tau_3 - \tau_1), \qquad (8.44)$$

where $g(\tau)$ can be the Gaussian function (Equation 8.14). The Gaussian window function has subtle difference from the Parzen window function (8.15), but produces a better approximation $\hat{m}_4^w(\tau_1, \tau_2, \tau_3)$ which we need.

Now, we can estimate the wavelet by matching the moment of a wavelet 'model' to the moment of wavelet $\hat{m}_4^w$ derived from seismic data. The objective function of the inverse problem is

$$\phi = \sum_{\tau_1, \tau_2, \tau_3} \| \hat{m}_4^w(\tau_1, \tau_2, \tau_3) - m_4^w(\tau_1, \tau_2, \tau_3) \|^2. \qquad (8.45)$$

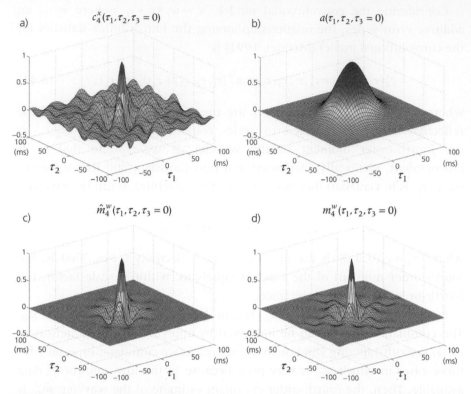

Figure 8.8 (*a*) A slice of the fourth-order cumulant, $c_4^x(\tau_1, \tau_2, \tau_3 = 0)$. (*b*) A slice of the 3-D Gaussian window $a(\tau_1, \tau_2, \tau_3 = 0)$. (*c*) A slice of the approximate fourth-order moment of the wavelet, $\hat{m}_4^w(\tau_1, \tau_2, \tau_3 = 0)$, which is a windowed fourth-order cumulant of the seismic trace. (*d*) A slice of the fourth-order moment of the wavelet $m_4^w(\tau_1, \tau_2, \tau_3 = 0)$.

Note that, in Equation 8.43, γ_4^r is a scalar, and $a(\tau_1, \tau_2, \tau_3)$ is a window that was pre-designed. Thus, this inverse problem is still referred to as cumulant matching.

Figure 8.8 displays four of the 2-D slices at $\tau_3 = 0$: a slice of the fourth-order cumulant, $c_4^x(\tau_1, \tau_2, \tau_3 = 0)$; a slice of the 3-D Gaussian window $a(\tau_1, \tau_2, \tau_3 = 0)$; a slice of the approximate fourth-order moment of the wavelet, $\hat{m}_4^w(\tau_1, \tau_2, \tau_3 = 0)$, which is a windowed fourth-order cumulant ofthe seismic trace; and a slice of the fourth-order moment of the wavelet $m_4^w(\tau_1, \tau_2, \tau_3 = 0)$.

Figure 8.9 displays a group of five seismic traces, and three estimated wavelets, which are the zero-phase wavelet, based on the amplitude

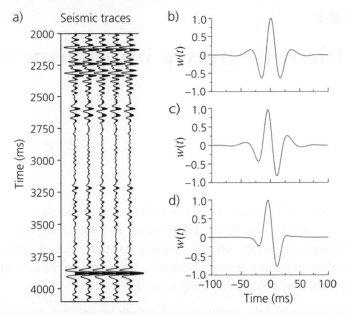

Figure 8.9 (*a*) A group of five seismic traces (zero mean). (*b*) Estimated zero-phase wavelet. (*c*) Estimated constant-phase wavelet. (*d*) Estimated mixed-phase wavelet using the cumulant matching inversion scheme.

spectrum, the constant-phase wavelet using the kurtosis matching method and the mixed-phase wavelet using the cumulant matching scheme.

The reliability of cumulant matching depends upon the quality of the cumulant estimate c_4^x calculated from noisy traces, the quality of the approximate moment $\hat{m}_4^w$ by a proper windowing, and the success in the minimisation of the multidimensional and nonlinear objective function.

CHAPTER 9

Seismic reflectivity inversion

Seismic reflectivity inversion retrieves reflectivity sequences from noise-contaminated seismic traces by eliminating the effect of the wavelet. Inverted reflectivity series directly represents the vertical variation of different sedimentary sequences. Generally speaking, however, there are two issues related with the reflectivity inversion:

1) Whether retrieved reflectivity series for each seismic trace is sparse and spiky;

2) Whether it is laterally coherent within a certain spatial range.

The term *sparse* refers to a solution that contains a few nonzero samples (reflectors), and *spiky* means a narrow waveform, which is close to an impulse that has a wide power spectrum in the frequency domain.

In the reflectivity inversion, it is appropriate to require that the model solution consists of the minimum number of events that can satisfy the data. Such a sparseness constraint, imposed on the unknown reflectivity model, also leads to a spiky solution consequentially. The method consists of solving the inverse problem with a regularisation, such as the L_1 norm and Cauchy criterion.

In order to improve spatial continuity of resultant reflectivity profiles, seismic reflectivity inversion shall be performed in a multichannel fashion.

9.1 The least-squares problem with a Gaussian constraint

In the Earth convolutional model, a seismic trace $\tilde{\mathbf{d}} = \{\tilde{d}_t\}$ that is discretised along the time axis t, can be expressed as

Seismic Inversion: Theory and Applications, First Edition. Yanghua Wang.
© 2017 John Wiley & Sons, Ltd. Published 2017 by John Wiley & Sons, Ltd.

$$\tilde{\mathbf{d}} = \mathbf{Gm} + \mathbf{e}, \tag{9.1}$$

where $\tilde{\mathbf{d}}$ is the data vector, $\mathbf{G}$ is the wavelet matrix in which each column contains the wavelet w_t properly padded with zeros in order to express discrete convolution, $\mathbf{m} = \{r_t\}$ is the vector of reflectivity series, and $\mathbf{e} = \{e_t\}$ is the additive data errors.

Assuming the wavelet w_t is pre-estimated through techniques presented in the previous chapter, the goal here is to recover a set of reflectivity series r_t from a recorded seismic trace $\tilde{d}_t$. The objective function is given as the following:

$$\phi(\mathbf{m}) = (\tilde{\mathbf{d}} - \mathbf{Gm})^{\mathrm{T}} \mathbf{C}_{\mathrm{d}}^{-1} (\tilde{\mathbf{d}} - \mathbf{Gm}) + \mu \mathbf{m}^{\mathrm{T}} \mathbf{C}_{\mathrm{m}}^{-1} \mathbf{m}, \tag{9.2}$$

where $\mathbf{C}_{\mathrm{d}}$ is the data covariance matrix, $\mathbf{C}_{\mathrm{m}}$ is the model covariance matrix, and μ balances the contribution between data fitting and model constraint. The solution is

$$\mathbf{m} = [\mathbf{G}^{\mathrm{T}} \mathbf{C}_{\mathrm{d}}^{-1} \mathbf{G} + \mu \mathbf{C}_{\mathrm{m}}^{-1}]^{-1} \mathbf{G}^{\mathrm{T}} \mathbf{C}_{\mathrm{d}}^{-1} \mathbf{d}. \tag{9.3}$$

This least-squares solution is a classical deconvolution, in which the factor μ, used to stabilise the solution, is called the pre-whitening factor. It is referred to as the percentage level of white noise in the data, and can be taken as a small fraction of the maximum value of the diagonal elements in $\mathbf{G}^{\mathrm{T}}\mathbf{G}$ (Robinson and Treitel, 1980). In practice, it is also treated as a damping parameter and is often tuned with an ad-hoc approach. Thus, this inversion formula is called widely as a damped least-squares method.

As stated in Chapter 6, minimising the objective function in Equation 9.2 is equivalent to maximising a conditional probability density function, in which both data errors and the model are assumed to have a Gaussian distribution:

$$p(\mathbf{d} \mid \mathbf{m}) = \frac{1}{(2\pi)^{M/2}\sqrt{\mid \mathbf{C}_{\mathrm{d}} \mid}} \exp\left(-\frac{1}{2}\mathbf{e}^{\mathrm{T}}\mathbf{C}_{\mathrm{d}}^{-1}\mathbf{e}^{\mathrm{T}}\right), \tag{9.4}$$

$$p(\mathbf{m}) = \frac{1}{(2\pi)^{N/2}\sqrt{\mid \mathbf{C}_{\mathrm{m}} \mid}} \exp\left(-\frac{1}{2}\mathbf{m}^{\mathrm{T}}\mathbf{C}_{\mathrm{m}}^{-1}\mathbf{m}\right), \tag{9.5}$$

where $\mid \mathbf{C}_{\mathrm{d}} \mid$ is the determinant of the data covariance matrix $\mathbf{C}_{\mathrm{d}}$, $\mid \mathbf{C}_{\mathrm{m}} \mid$ is the determinant of the model covariance matrix, and M is the number of data samples, and N is the number of random variables. The pre-multiplying factor is simply a normalisation so that the integral of the probability is unity.

Equation 9.4 assumes that the additive error series **e** is generally in Gaussian distribution. Such an assumption is appropriate in most situations in seismic inversion.

The physical meaning of the Gaussian model constraint, Equation 9.5, is that the inversion will not allow a significant fraction of 'outliers' to exist in the solution. These outliers are values that lie many standard deviations away from the mean. Equation 9.5 indicates that the value of the Gaussian distribution is practically zero when the value m_k lies more than a few standard deviations away from the mean.

The effect of this Gaussian model constraint (L$_p$ norm with $p = 2$) in seismic reflectivity inversion is demonstrated in Figure 9.1. Whereas Figure 9.1a is an original seismic section, Figure 9.1b shows neither sparse nor spiky behaviour of the inversion with the L$_2$ norm constraint. This undesirable feature suggests that, for seismic reflectivity inversion, an appropriate constraint would be a probability density function with a heavy-tailed distribution.

9.2 Reflectivity inversion with an L$_p$ norm constraint

In order to produce a solution estimate that is sparse and spiky enough to be a reflectivity series, an L$_p$ norm model constraint with $p \leq 1$ may be used to supersede the L$_2$ norm constraint.

The objective function with an L$_p$ norm constraint is defined as

$$\phi(\mathbf{m}) = (\tilde{\mathbf{d}} - \mathbf{Gm})^{\mathrm{T}} \mathbf{C}_{\mathrm{d}}^{-1} (\tilde{\mathbf{d}} - \mathbf{Gm}) + \mu (\hat{\mathbf{m}}^{p/2})^{\mathrm{T}} \mathbf{C}_{\mathrm{m}}^{-1} \hat{\mathbf{m}}^{p/2}, \tag{9.6}$$

where $\hat{\mathbf{m}}^{p/2}$ is a vector with the element $|m_k|^{p/2}$, and $(\hat{\mathbf{m}}^{p/2})^{\mathrm{T}} \hat{\mathbf{m}}^{p/2}$ $= \| \mathbf{m} \|_p^p$. Hence, this is a generalisation of the objective function (9.2), which uses the L$_2$ model constraint. The L$_p$ norm constraint in Equation 9.6 is equivalent to the probability density function

$$p(\mathbf{m}) \propto \exp\left(-\frac{1}{2} (\hat{\mathbf{m}}^{p/2})^{\mathrm{T}} \mathbf{C}_{\mathrm{m}}^{-1} \hat{\mathbf{m}}^{p/2} \right). \tag{9.7}$$

Note that $\mathbf{C}_{\mathrm{m}}$ has only a physical interpretation for $p = 2$ (Johnson and Kotz, 1972). But, we can simply treat $\mathbf{C}_{\mathrm{m}}$ as a weighting matrix that can also cancel the physical units in $\| \mathbf{m} \|_p^p$.

The L$_p$ norm function has a different statistical meaning when p has varied values. When p equals ∞, 2 and 1, the solutions of L$_p$ norm

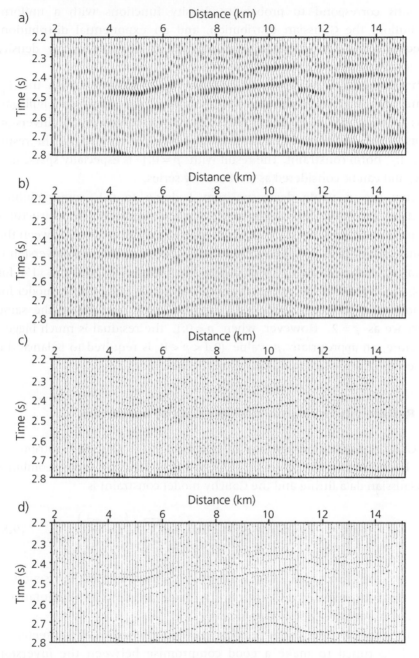

Figure 9.1 (*a*) A seismic section. (*b*) Reflectivity solution of the L_2 norm method. (*c*) Reflectivity solution of the L_1 norm method. (*d*) Reflectivity solution of the L_p norm method with $p = 0.1$. This inversion solution is sparse and spiky enough to be the reflectivity series.

functions correspond to probability density functions with a uniform distribution, the Gaussian distribution, and an exponential distribution, respectively. As the p value approaches zero, the probability density function becomes more similar to an impulse.

Minimisation of the objective function in Equation 9.6 with the L_p norm, $p \le 1$, favours a solution with isolated spikes (Levy and Fullagar, 1981; Debeye and van Riel, 1990). Figure 9.1 compares the inversion results of the L_p norm model constraint, for $p = 1$ and 0.1, to the result of an L_2 norm constraint. The result with $p = 0.1$ is especially sparse and spiky, and can be considered as a reflectivity series.

When the p value decreases from 1 down to 0.1, some minimal reflections are muted (sparseness), whereas the major reflection structures are emphasised (spikiness). As the consequence, the difference between the original seismic data and the synthetic data generated from the inversion is increased. In this example, the data differences are 1.6%, 1.4% and 11%, for $p = 2$, 1, and 0.1, respectively. The result of $p = 2$ is the initial model for the following inversion with $p < 2$, and so $p = 1$ has almost the same difference as $p = 2$. However, when $p = 0.1$, the residual is much bigger. Therefore, an appropriate p value $(0.1 < p < 1)$ is required to balance the sparseness of the solution estimate and the data residual.

9.3 Reflectivity inversion with the Cauchy constraint

The Cauchy distribution also has fatter tails than the Gaussian distribution (the L_2 norm model constraint). The objective function which combines the Gaussian data fitting and the Cauchy model constraint is

$$\phi(\mathbf{m}) = \frac{1}{2}(\tilde{\mathbf{d}} - \mathbf{Gm})^{\mathrm{T}} \mathbf{C}_{\mathrm{d}}^{-1}(\tilde{\mathbf{d}} - \mathbf{Gm}) + \mu \sum_{k=1}^{N} \ln\left(1 + \frac{m_k^2}{\lambda^2}\right). \tag{9.8}$$

The Cauchy parameter λ plays a key role in controlling the sparseness of the reflectivity inversion. Chapter 6 has made an effort to quantitatively estimate an optimal Cauchy parameter λ.

In the same manner as the previous method, the trade-off parameter μ should be tuned to make a good compromise between the inversion sparseness and the data residual. That is, the influence of the Cauchy regularisation needs to be weakened in the objective function. Otherwise, the result would be too sparse, as reflectivity samples with small amplitudes

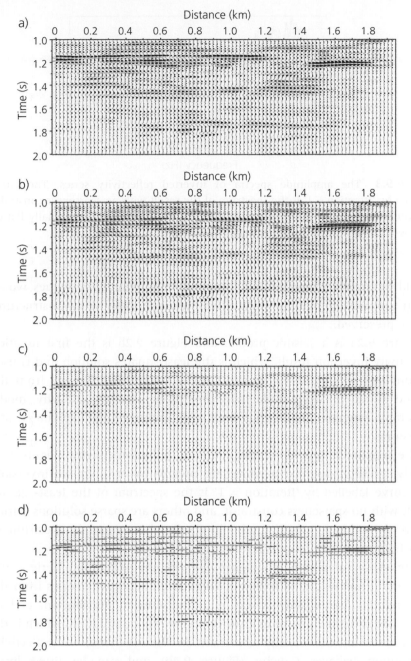

Figure 9.2 Seismic reflectivity inversion with Cauchy constraint. (*a*) A seismic profile. (*b*) The first iteration result, which is a least-squares solution. (*c*) The inversion results after the second iteration. (*d*) The inversion results after the eighth iteration.

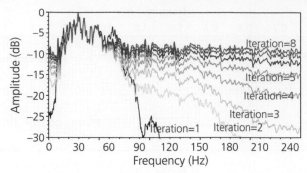

Figure 9.3 The amplitude spectra of inverted reflectivity series. The curve labelled by 'Iteration = 1' is the spectrum of the result with no sparseness constraint and others are sparse solutions during different iterations, which gradually flatten the spectrum with more high-frequency components in reflectivity series.

would be largely eliminated. Eliminating small reflectivity values would mean suppressing minor geological features, although the main structures can be preserved.

Figure 9.2a is a seismic profile, and Figure 9.2b is the first iteration reflectivity inversion result, which is the least-squares solution and is used for quantitative estimation of the Cauchy parameter. Starting from the second iteration, using reflectivity inversion with the Cauchy model constraint, iterative inversion gradually improves the spikiness and sparseness.

The amplitude spectra of the resultant reflectivity series (Figure 9.3) clearly indicate progressive whitening process in the iterative inversion. The curve labelled by 'Iteration = 1' is the spectrum of the least-squares result with no sparseness constraint, and others are sparse solutions during different iterations, which gradually flatten the amplitude spectrum of high-frequency components in reflectivity series.

The statistical summary (Figure 9.4) of the reflectivity series after different iterations clearly indicates that seismic reflectivity series from the conventional least-squares inversion is in Gaussian distribution, and all others are more likely to be in Cauchy. In the second iteration, which is the first time using the Cauchy constraint in the inversion, the reflectivity distribution behaves Cauchy (Figure 9.4b) and can be fitted by a continuous curve with Cauchy parameter $\lambda = 1.74 \times 10^{-3}$.

However, after further iterations, the distribution gradually changes towards a Gaussian distribution. In the eighth iteration (Figure 9.4c) the

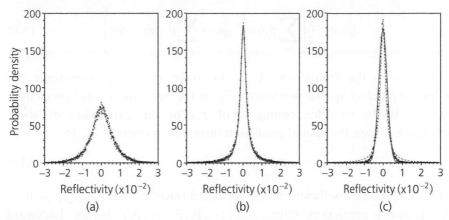

Figure 9.4 The statistical information of seismic reflectivity after a different number of iterations. (*a*) Reflectivity distribution after the first iteration, which is in Gaussian distribution. (*b*) Cauchy fitting of the reflectivity distribution after the second iteration (the first Cauchy constrained inversion). (*c*) After the eighth iteration, the reflectivity distribution changes towards Gaussian (solid line) rather than Cauchy (dashed line).

reflectivity series is fitted by the Gaussian function with standard deviation $\sigma = 2.24 \times 10^{-3}$ (a solid curve). For comparison, the dashed line is the Cauchy distribution.

This example reveals that the least-squares solution with a Cauchy regularisation results in a Gaussian distribution that is very close to a Cauchy distribution.

9.4 Multichannel reflectivity inversion

As seismic reflectivity inversion, with either the L_p norm ($p \le 1$) constraint or the Cauchy constraint, is performed on a trace-by-trace basis, it may lead to lateral discontinuity on the resultant reflectivity profiles, especially in the case where seismic data for complicated structures have a low signal-to-noise ratio. Therefore, we propose a multichannel inversion scheme, for improving the lateral coherency of the reflectivity profile and for a clear structure characterisation.

In the multichannel inversion method, the information of adjacent traces is incorporated by spatial prediction, during each iteration of the inversion process. We assume that seismic reflectivity series is spatially predictable in the frequency domain:

$$\tilde{r}_k(\omega) = \frac{1}{2}\left(\sum_{j=1}^{J} p_j(\omega)r_{k-j}(\omega) + \sum_{j=1}^{J} \overline{p}_j(\omega)r_{k+j}(\omega)\right), \qquad (9.9)$$

where ω is the frequency, k is the trace index, r_k represents the reflectivity before spatial prediction, $\tilde{r}_k$ is the reflectivity after prediction, and $\overline{p}_j$ is the complex conjugate of p_j. In the expression of $\omega - x$ prediction above, the spatial prediction filter, for frequency ω, is

$$p_J, \ \cdots, \ p_2, \ p_1, \ -2, \ \overline{p}_1, \ \overline{p}_2, \ \cdots, \ \overline{p}_J, \qquad (9.10)$$

where -2 is the coefficient of the predicted trace $\tilde{r}_k$, $\{p_J, \cdots, p_2, p_1, -1\}$ is the forward prediction filter, and $\{-1, \overline{p}_1, \overline{p}_2, \cdots, \overline{p}_J\}$ is the backward prediction filter.

For a 2-D data slice (for 3-D seismic data application), the shape of the prediction filter has the form

$$\begin{bmatrix} p_{2,2} & p_{2,1} & p_{2,0} \\ p_{1,2} & p_{1,1} & p_{1,0} \\ p_{0,2} & p_{0,1} & -1 \\ p_{-1,2} & p_{-1,1} \\ p_{-2,2} & p_{-2,1} \end{bmatrix} \quad \text{and} \quad \begin{bmatrix} & \overline{p}_{-2,1} & \overline{p}_{-2,2} \\ & \overline{p}_{-1,1} & \overline{p}_{-1,2} \\ -1 & \overline{p}_{0,1} & \overline{p}_{0,2} \\ \overline{p}_{1,0} & \overline{p}_{1,1} & \overline{p}_{1,2} \\ \overline{p}_{2,0} & \overline{p}_{2,1} & \overline{p}_{2,2} \end{bmatrix} \qquad (9.11)$$

for the forward and backward predictions, respectively (Wang, 1999b, 2002). In this case, where the size of the prediction operator is 5×5, for example, the output position is under the '-1' coefficients on the forward and backward prediction operators.

Finding the prediction filter itself is an inverse problem. For example, for a 1-D data series (for 2-D seismic data application), we rewrite Equation 9.9 into forward and backward predictions separately as

$$r_k(\omega) = \sum_{j=1}^{J} p_j(\omega)r_{k-j}(\omega),$$

$$\tilde{r}_k(\omega) = \sum_{j=1}^{J} p_j(\omega)\overline{r}_{k+j}(\omega). \qquad (9.12)$$

These two equations are spatial convolution, for a given frequency ω. Thus, the prediction procedure is often called spatial deconvolution. We can express Equation 9.12 in the matrix-vector form, $\mathbf{Ap} = \mathbf{b}$, where $\mathbf{A}$ is a matrix consisting of elements r_{k-j} and $\bar{r}_{k+j}$, $\mathbf{p}$ is the prediction filter $\{p_j\}$, and

$$\mathbf{b} = \begin{bmatrix} \mathbf{r} \\ \bar{\mathbf{r}} \end{bmatrix}. \tag{9.13}$$

We may obtain the filter $\mathbf{p}$, using a stabilised least-squares method,

$$\mathbf{p} = [\mathbf{A}^H \mathbf{A} + \mu \mathbf{I}]^{-1} \mathbf{A}^H \mathbf{b}, \tag{9.14}$$

where the superscript ^{H} denotes the Hermitian transpose, the transpose of the complex conjugate. Recall that filter $\mathbf{p}$ is constructed in Equation 9.12, based on data samples in both sides. Therefore, coefficients of this two-sided prediction filter in Equation 9.10 have conjugate symmetry, and the prediction filter is zero phased in spatial frequency space.

Let us present a reflectivity profile as a data matrix, $\mathbf{R}(t, x)$, in which each column is a reflectivity trace $r(t)$, and x is the spatial position. Its frequency domain counterpart can be obtained by

$$\mathbf{R}(\omega, x) = \mathbf{F}\mathbf{R}(t, x), \tag{9.15}$$

where $\mathbf{F}$ denotes the 1-D Fourier transform, with respect to the time t.

Spatial prediction in the frequency domain can be expressed as

$$\tilde{\mathbf{R}}(\omega, x) = P\{\mathbf{F}\mathbf{R}(t, x)\}, \tag{9.16}$$

where P is the generalised prediction operation. The inverse Fourier transform will generate a time-domain data matrix:

$$\tilde{\mathbf{R}}(t, x) = \mathbf{F}^{-1}[P\{\mathbf{F}\mathbf{R}(t, x)\}]. \tag{9.17}$$

Let us summarise the entire prediction processing, including the processes of forward and inverse Fourier transform and the frequency-domain spatial prediction in between, using a single operator $\mathbf{L}$, so that

$$\tilde{\mathbf{R}}(t, x) = \mathbf{L}\mathbf{R}(t, x). \tag{9.18}$$

Because we know the data matrices $\mathbf{R}(t, x)$ and $\tilde{\mathbf{R}}(t, x)$, we are able to calculate explicitly the spatial prediction operator $\mathbf{L}$ by

$$\mathbf{L} = \tilde{\mathbf{R}}\mathbf{R}^T (\mathbf{R}\mathbf{R}^T + \mu \mathbf{I})^{-1}. \tag{9.19}$$

This is a least-squares solution with stabilisation factor μ.

Convoluting reflectivities with a wavelet, we have a seismic trace matrix:

$$\mathbf{S} = \mathbf{WLR} = \mathbf{GR} , \qquad (9.20)$$

where $\mathbf{G} = \mathbf{WL}$. Therefore, the reflectivity series using the multichannel Cauchy inversion can be expressed as

$$\mathbf{R} = \left(\mathbf{G}^{\mathrm{T}} \mathbf{C}_{\mathrm{d}}^{-1} \mathbf{G} + \frac{\mu}{\lambda^2} \mathbf{D} \right)^{-1} \mathbf{G}^{\mathrm{T}} \mathbf{C}_{\mathrm{d}}^{-1} \tilde{\mathbf{S}} , \qquad (9.21)$$

where $\mathbf{D}$ is a modified diagonal matrix from Equation 6.45, related to the Cauchy constraint, and $\tilde{\mathbf{S}}$ is the observed seismic section. Thus, Equation 9.21 is a simple extension of the Cauchy solution, given by Equation 6.44, to the multichannel case.

9.5 Multichannel conjugate gradient method

For the implementation of multichannel reflectivity inversion, we can modify the conjugate gradient method, which conventionally is applicable to single channel inversion, to a multichannel fashion.

Initially, a reflectivity series is obtained by minimising $\| \tilde{\mathbf{s}} - \mathbf{Wr} \|^2$ trace by trace, where $\tilde{\mathbf{s}}$ is a seismic trace, $\mathbf{W}$ is the wavelet matrix, $\mathbf{r}$ is the reflectivity series, and $\mathbf{Wr}$ is the discrete convolution in the time domain. The reflectivity series is obtained by

$$\mathbf{r} = (\mathbf{W}^{\mathrm{T}} \mathbf{W} + \mu \mathbf{I})^{-1} \mathbf{W}^{\mathrm{T}} \tilde{\mathbf{s}} . \qquad (9.22)$$

This is a standard seismic deconvolution (Equation 9.3). A group of reflectivity traces makes a data matrix $\mathbf{R}(t,x) = [\mathbf{r}_1, \ \mathbf{r}_2, \cdots, \ \mathbf{r}_N]$, where N is the number of traces.

Instead of a direct solution, which involves the calculation of matrix inverse, we can re-write Equation 9.21 as

$$\left(\mathbf{G}^{\mathrm{T}} \mathbf{C}_{\mathrm{d}}^{-1} \mathbf{G} + \frac{\mu}{\lambda^2} \mathbf{D} \right) \mathbf{R} = \mathbf{G}^{\mathrm{T}} \mathbf{C}_{\mathrm{d}}^{-1} \tilde{\mathbf{S}} . \qquad (9.23)$$

The difference between the two sides is

$$\mathbf{E} = \mathbf{G}^{\mathrm{T}} \mathbf{C}_{\mathrm{d}}^{-1} \tilde{\mathbf{S}} - \left(\mathbf{G}^{\mathrm{T}} \mathbf{C}_{\mathrm{d}}^{-1} \mathbf{G} + \frac{\mu}{\lambda^2} \mathbf{D} \right) \mathbf{R} . \qquad (9.24)$$

Using this residual matrix, we can extend the conventional conjugate gradient method in a multichannel fashion.

We can design the multichannel inversion procedure as what follows.

Step 1: initialise the value of $\mathbf{R}^{(1)} = [\mathbf{r}_1^{(1)}, \mathbf{r}_2^{(1)}, \cdots, \mathbf{r}_N^{(1)}]$, set the residual $\mathbf{E}^{(1)} = [\mathbf{e}_1^{(1)}, \mathbf{e}_2^{(1)}, \cdots, \mathbf{e}_N^{(1)}]$, and set the search direction $\mathbf{P}^{(1)} = [\mathbf{p}_1^{(1)}, \mathbf{p}_2^{(1)}, \cdots, \mathbf{p}_N^{(1)}]$ equal to $\mathbf{E}^{(1)}$.

Step 2: implement the conjugate gradient iteration, for $k = 1, 2, \cdots$:
calculate the matrix

$$\mathbf{A} = \mathbf{G}^{\mathrm{T}}\mathbf{C}_{\mathrm{d}}^{-1}\mathbf{G} + \frac{\mu}{\lambda^2}\mathbf{D}, \qquad (9.25)$$

and execute the inner loop, for $j = 1, 2, \cdots, N$,

$$\mathbf{u}_j^{(k)} = \mathbf{A}\mathbf{p}_j^{(k)}, \qquad\qquad \alpha_j^{(k)} = \frac{(\mathbf{e}_j^{(k)}, \mathbf{e}_j^{(k)})}{(\mathbf{p}_j^{(k)}, \mathbf{u}_j^{(k)})},$$

$$\mathbf{r}_j^{(k+1)} = \mathbf{r}_j^{(k)} + \alpha_j^{(k)}\mathbf{p}_j^{(k)}, \qquad \mathbf{e}_j^{(k+1)} = \mathbf{e}_j^{(k)} - \alpha_j^{(k)}\mathbf{u}_j^{(k)}, \qquad (9.26)$$

$$\beta_j^{(k)} = \frac{(\mathbf{e}_j^{(k+1)}, \mathbf{e}_j^{(k+1)})}{(\mathbf{e}_j^{(k)}, \mathbf{e}_j^{(k)})}, \qquad \mathbf{p}_j^{(k+1)} = \mathbf{e}_j^{(k+1)} + \beta_j^{(k)}\mathbf{p}_j^{(k)},$$

where $\mathbf{u}_j$, $\mathbf{p}_j$, $\mathbf{e}_j$ and $\mathbf{r}_j$ represent single vectors corresponding to each trace j, and $\alpha_j^{(k)}$ and $\beta_j^{(k)}$ are the kth step length and the conjugate coefficient, respectively.

Step 3: reinitialise $\mathbf{R} = [\mathbf{r}_1, \mathbf{r}_2, \cdots, \mathbf{r}_N]$, using the preliminary result obtained from the conjugate gradient method in Step 2, and restart the multichannel process from Step 1.

Step 4: calculate the reflectivity series, finally, using $\tilde{\mathbf{R}} = \mathbf{L}\mathbf{R}$.

In this multichannel implementation, once the residual of any trace is under a predefined threshold, the reflectivity values of this trace are stored in the data matrix $\mathbf{R} = [\mathbf{r}_1, \mathbf{r}_2, \cdots, \mathbf{r}_N]$, whereas other traces keep updating through further iterations. The conjugate gradient process will stop, when the residuals of all traces are below the threshold.

In the multichannel conjugate gradient method, the operator $\mathbf{A}$ is a multichannel operator. However, each inside iteration loop involves the calculation of a single channel conjugate gradient method, such as the inner production of vectors and summation of vectors, etc.

Once a satisfactory result is obtained through repeating the processes in the first three steps, there is a fourth and final step, Equation 9.18.

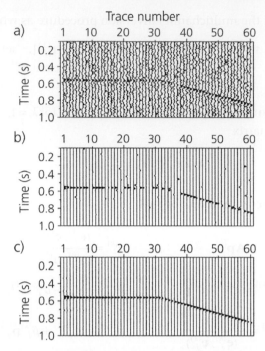

Figure 9.5   (*a*) A synthetic seismic profile. (*b*) The reflectivity profile obtained by the standard single channel deconvolution. (*c*) The reflectivity profile obtained by the multichannel inversion algorithm.

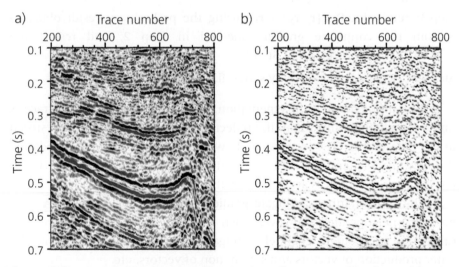

Figure 9.6   (*a*) A field seismic profile. (*b*) The reflectivity profile obtained by the multichannel reflectivity inversion method.

Figure 9.5 clearly demonstrates the effectiveness of the multichannel inversion algorithm. When applying the coherence information of the adjacent traces in the reflectivity inversion, lateral continuity has been well-preserved.

Figure 9.6a is a field seismic profile, and Figure 9.6b is the inversion result, obtained by the four-step, multichannel reflectivity inversion method.

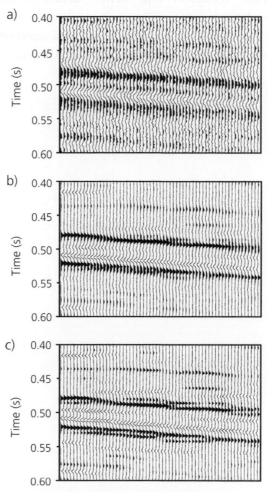

Figure 9.7 (*a*) A zoomed-in field seismic profile. (*b*) The reflectivity profile obtained by the single channel inversion, followed by spatial prediction. (*c*) The reflectivity profile obtained by multichannel reflectivity inversion (the first three steps), followed by a spatial prediction (the fourth and final step).

Figure 9.7 displays zoomed-in profiles. Figure 9.7a is the zoomed-in field seismic profile, which contains a reasonably-low level of noise. Figure 9.7b illustrates the conventional single channel inversion, in which the spatial continuity has been improved by $\omega - x$ prediction filtering, applied to the result. This spatial prediction filter can smooth the reflectivity.

Figure 9.7c is the reflectivity profile obtained by the multichannel reflectivity inversion scheme (the first three steps), followed by a spatial prediction, $\tilde{\mathbf{R}} = \mathbf{LR}$, the fourth and final step. This final reflectivity profile has a better noise reduction and better structural characterisation, compared to any others.

In summary, there are two key elements for a successful reflectivity inversion: sparseness constraint, and multichannel scheme.

CHAPTER 10

Seismic ray-impedance inversion

Seismic impedance is widely recognised as a powerful lithological indicator for discriminating hydrocarbon reservoirs. Ideally, once the reflectivity series is obtained, the impedance variation can be calculated by a recursive formula. In order to mitigate the accumulative errors due to recursion and to stabilise the computation, the impedance variation should be estimated by inversion, rather than a direct calculation, incorporating any geological constraints, such as well logging and borehole seismic information, in the least-squares solution.

There are different types of impedance quantities that can be inverted from seismic data: the acoustic impedance, the elastic impedance and the ray impedance. While the acoustic impedance neglects the rigidity of the subsurface media, the elastic impedance takes account of the shear modules. However, both types of impedance parameters violate Snell's law, when the inversion is performed on nonzero offset traces. The ray impedance concept is an extension of the acoustic and elastic impedance, and is an impedance quantity measured along a ray path, which honours the real physics by taking account of the non-normal incident angles correctly (Wang, 2003b). Thus, this chapter will present only the inversion strategy for the ray impedance inversion.

10.1 Acoustic and elastic impedances

Acoustic impedance (*AI*) is the product of *P*-wave velocity α and bulk density ρ:

$$AI = \rho\alpha . \tag{10.1}$$

Seismic Inversion: Theory and Applications, First Edition. Yanghua Wang.
© 2017 John Wiley & Sons, Ltd. Published 2017 by John Wiley & Sons, Ltd.

In seismology, the term 'acoustic media' means the media that do not support shear wave propagation or simply have a zero-valued S-wave velocity. Hence, the reflectivity is related only to the P-wave velocity α and bulk density ρ. A reflection coefficient r_k of a normal incidence at a subsurface interface is given by

$$r_k = \frac{AI_{k+1} - AI_k}{AI_{k+1} + AI_k},$$ (10.2)

where AI_k and AI_{k+1} are, respectively, the acoustic impedances at the wave incident layer (k) and at the opposite side $(k+1)$ of the reflecting interface. Then the acoustic impedance ratio is

$$\frac{AI_{k+1}}{AI_k} = \frac{1+r_k}{1-r_k}.$$ (10.3)

Taking two approximations $1+r_k \approx \exp(r_k)$, and $1-r_k \approx \exp(-r_k)$, this ratio can be expressed also as

$$\frac{AI_{k+1}}{AI_k} = \exp(2r_k).$$ (10.4)

The Taylor expansions of Equations 10.3 and 10.4 for the impedance ratio are

$$\frac{1+r_k}{1-r_k} = 1 + 2\sum_{n=1}^{\infty} r_k^n = 1 + 2r_k + 2r_k^2 + 2r_k^3 + \cdots,$$ (10.5)

and

$$\exp(2r_k) = \sum_{n=0}^{\infty} \frac{2^n}{n!} r_k^n = 1 + 2r_k + 2r_k^2 + \frac{4}{3}r_k^3 + \cdots.$$ (10.6)

This comparison indicates that the formulae are equivalent up to second order in r_k. As shown in Figure 10.1, for $|r_k| < 0.4$, the difference between these two formulae is less than 5%. Theoretically, the absolute value of a reflection coefficient is usually less than 0.4. However, if considering the data noise and possible processing errors, the exponential formula would be more stable in the impedance computation than the other formula. For multi-layered media, Equations 10.3 and 10.4 become

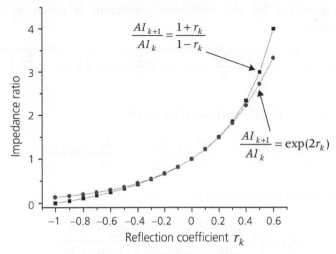

Figure 10.1 Difference between two formulae for the acoustic impedance ratio.

$$AI_{k+1} = AI_1 \prod_{i=1}^{k} \frac{1+r_i}{1-r_i} = AI_1 \exp\left(2\sum_{i=1}^{k} r_i\right). \qquad (10.7)$$

These expressions relate the acoustic impedance AI_{k+1} to the reflection coefficients $\{r_i\}$. Considering possible errors $\{e_i\}$ in 'data' $\{\tilde{r}_i\}$, we have

$$AI_{k+1} = AI_1 \exp\left(2\sum_{i=1}^{k} \tilde{r}_i\right) = AI_1 \exp\left(2\sum_{i=1}^{k} r_i\right)\exp\left(2\sum_{i=1}^{k} e_i\right). \qquad (10.8)$$

The second exponential function $\exp(2\sum_i e_i) \to 1$, if assuming $\{e_i\}$ are random errors. This proves that the exponential expression for the impedance ratio would be more stable than the production formula in the impedance computation.

Elastic impedance (*EI*) is a generalisation of the acoustic impedance. It combines the *S*-wave velocity with the *P*-wave velocity and density to construct an indicator for discriminating hydrocarbon reservoirs.

For defining the *EI* parameter, a reflection coefficient can be rewritten in a formula analogous to the acoustic reflectivity as

$$r_k(\theta) = \frac{EI_{k+1}(\theta) - EI_k(\theta)}{EI_{k+1}(\theta) + EI_k(\theta)} \approx \frac{1}{2}\ln\frac{EI_{k+1}(\theta)}{EI_k(\theta)}. \qquad (10.9)$$

A linear expression for the reflection coefficient is given by Aki and Richards (1980):

$$r(\theta) = \frac{1}{2}\left((1 + \tan^2 \theta)\frac{\Delta\alpha}{\alpha} - 8K\sin^2\theta\frac{\Delta\beta}{\beta} + (1 - 4K\sin^2\theta)\frac{\Delta\rho}{\rho} \right),$$

(10.10)

where $K = \beta^2/\alpha^2$. For the kth interface, with

$$\frac{\Delta\alpha}{\alpha} \approx \ln\frac{\alpha_{k+1}}{\alpha_k}, \qquad \frac{\Delta\beta}{\beta} \approx \ln\frac{\beta_{k+1}}{\beta_k}, \qquad \frac{\Delta\rho}{\rho} \approx \ln\frac{\rho_{k+1}}{\rho_k},$$

(10.11)

Equation 10.10 can be expressed as

$$r_k(\theta_k) \approx \frac{1}{2}\ln\frac{\alpha_{k+1}^{1+\tan^2\theta_k}\beta_{k+1}^{-8K_k\sin^2\theta_k}\rho_{k+1}^{1-4K_k\sin^2\theta_k}}{\alpha_k^{1+\tan^2\theta_k}\beta_k^{-8K_k\sin^2\theta_k}\rho_k^{1-4K_k\sin^2\theta_k}}.$$

(10.12)

If assuming $K_k = K$ and $\theta_k = \theta$, both K and θ being constant, the above expression becomes the recursive formula (10.9). Then, the *EI* parameter is defined as

$$EI_k(\theta) = \rho_k\alpha_k(\alpha_k^{\tan^2\theta}\beta_k^{-8K\sin^2\theta}\rho_k^{-4K\sin^2\theta}).$$

(10.13)

That is, the *EI* parameter is the acoustic impedance $\rho_k\alpha_k$ multiplied by a factor $\alpha_k^{\tan^2\theta}\beta_k^{-8K\sin^2\theta}\rho_k^{-4K\sin^2\theta}$, which takes into account the *S*-wave velocity and the wave incident angle.

Since *EI* is defined with respect to variable incidence angle (Connolly, 1999), *EI* inversion provides a consistent framework to invert nonzero offset seismic data, just as the *AI* inversion does for zero-offset data.

However, there are two assumptions involved in the derivation that make the *EI* parameter inaccurate. First, the incident and transmitted angles are the same value, and secondly the velocity ratio is a constant. In fact, the incident and transmitted angles should follow Snell's law, and the *P*- and *S*-wave velocity ratio in most cases will change at different sides of the interface. Improvement leads to a ray-impedance concept.

10.2 Ray impedance

Ray impedance (*RI*) defines the elastic impedance along a ray path with constant ray parameter *p*, and hence honours Snell's law (Wang, 2003b).

The reflection coefficient at an interface depends on the relative change in ray impedances across the interface:

$$r_k(p) = \frac{RI_{k+1}(p) - RI_k(p)}{RI_{k+1}(p) + RI_k(p)} \approx \frac{1}{2} \ln \frac{RI_{k+1}(p)}{RI_k(p)}. \tag{10.14}$$

The derivation is based on the following quadratic formula of the reflection coefficient (Wang, 1999a)

$$r(p) = \frac{\rho_2 q_{\alpha_1} - \rho_1 q_{\alpha_2}}{\rho_2 q_{\alpha_1} + \rho_1 q_{\alpha_2}} - \frac{2\Delta\mu}{\rho} p^2, \tag{10.15}$$

where $q_{\alpha_1} = (1/\alpha_1^2 - p^2)^{1/2}$ and $q_{\alpha_2} = (1/\alpha_2^2 - p^2)^{1/2}$ are the P-wave vertical slownesses, ρ_1 and ρ_2 are bulk densities, ρ is their average, and $\mu = \rho\beta^2$ is the shear modulus. This expression may be written in a recursive form as

$$r_k(\theta_k, \theta_{k+1}) \approx \frac{1}{2} \ln \frac{\rho_{k+1}\alpha_{k+1} \dfrac{\cos^{4(\eta+2)} \varphi_{k+1}}{\cos\theta_{k+1}}}{\rho_k\alpha_k \dfrac{\cos^{4(\eta+2)} \varphi_k}{\cos\theta_k}}, \tag{10.16}$$

where φ_k and φ_{k+1} are the PS-wave reflection and transmission angles, respectively, and

$$\eta = \frac{\Delta\rho/\rho}{\Delta\beta/\beta}. \tag{10.17}$$

This recursive expression leads to the definitions of the ray impedance:

$$RI_k(\theta_k) = \frac{\rho_k\alpha_k}{\cos\theta_k} \left(1 - \frac{\beta_k^2}{\alpha_k^2} \sin^2\theta_k \right)^{2(\eta+2)}, \tag{10.18}$$

and

$$RI_k(p) = \frac{\rho_k\alpha_k}{\sqrt{1 - \alpha_k^2 p^2}} (1 - \beta_k^2 p^2)^{2(\eta+2)}. \tag{10.19}$$

The assumption here is that η is a constant. It is justified, as $\Delta\rho/\rho$ and $\Delta\beta/\beta$ generally have a high covariance.

Note also that the factor

$$AI_k(p \text{ or } \theta_k) = \frac{\rho_k\alpha_k}{\sqrt{1 - \alpha_k^2 p^2}} = \frac{\rho_k\alpha_k}{\cos\theta_k} \tag{10.20}$$

is the definition of the acoustic impedance along a ray path p or with incident angle θ. With this concept, an acoustic impedance inversion can be realised also in the pre-stack domain.

Figure 10.2 displays two cross-plots of ray-impedance at $p = 150$ ms/km and $p = 0$. These data are from a clastic reservoir between the depths 2200 and 2330 m. First, we can use cross-plot to discriminate the sand

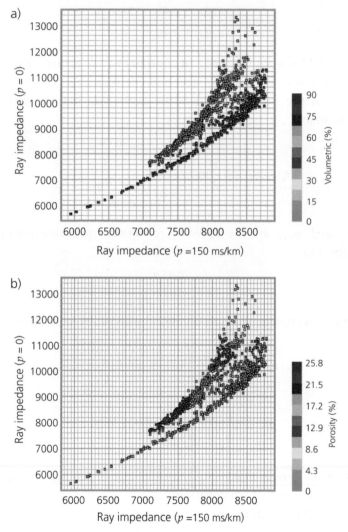

Figure 10.2 Cross-plots of the ray impedance at $p = 150$ ms/km and that at $p = 0$, for (*a*) revealing the shale content (%) within a target reservoir at depth 2200– 2300 m, and (*b*) the porosity (%) of different sand bodies.

from shale, based on shale content, as shown in Figure 10.2a. Then, we can further discern the sand bodies with different porosity, as shown in Figure 10.2b.

The ray impedance concept is an extension of the conventional acoustic impedance and the elastic impedance, but has a natural advantage in seismic inversion since it is defined on a specific ray parameter p, in which case both the *PP*- and *PS*-wave reflections honour Snell's law (Figure 10.3). The ray impedances obtained from *PP*- and *PS*-wave reflection data can be used jointly in an inversion, to invert for the elastic parameters, such as *P*- and *S*-wave velocities, the bulk density, and the Lamé parameters, etc.

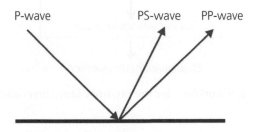

P-wave   PS-wave   PP-wave

Figure 10.3 The *PP*- and *PS*-wave reflections with a constant ray-parameter (p) share the same reflection point. Therefore, the *PP*-wave and *PS*-wave data can be used jointly for the elastic parameters.

In summary, the ray impedance concept shows, at least, the following three advancements:

1) It relaxes the assumptions in the *EI* definition, so that *RI* follows true physics, i.e. Snell's law, and the *S*- to *P*-wave velocity ratio varies following true geology (variable at different depths).

2) It has a potential advantage in lithology identification.

3) It makes a solid base for a joint inversion of *PP* and *PS*-waves, as they share the same constant ray-parameter p.

10.3 Workflow of ray-impedance inversion

Ray impedance is a lithological parameter that measures the elastic impedances along a specific ray path with a constant ray parameter p.

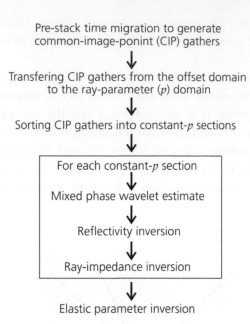

Figure 10.4 A typical workflow for the ray-impedance inversion.

Figure 10.4 presents a typical workflow for the ray-impedance inversion in the pre-stack domain.

For the ray impedance inversion, we first implement seismic migration in the pre-stack domain, to produce common-reflection-point (CRP) gathers. As an example, shown in Figure 10.5, a CRP gather is often in the source-receiver offset domain, and needs to be transferred to the ray-parameter (p) domain.

The CRP gathers are sorted into constant-p sections (Figure 10.6). For each constant-p section, we perform the mixed-phase wavelet estimation, the sparse reflectivity inversion, and the ray impedance inversion. Those obtained ray impedances from different constant-p sections can be used jointly to invert further for the elastic parameters, such as the P-wave velocity, the S-wave velocity, and the density.

10.4 Reflectivity inversion in the ray-parameter domain

In the ray-parameter domain, reflectivity inversion may be subject to some geological constraints, such as the impedance model interpolated from

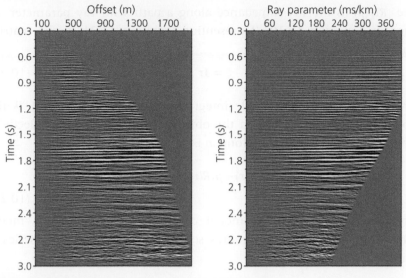

Figure 10.5 A common-reflection-point (CRP) gather in the offset domain and in the ray-parameter domain.

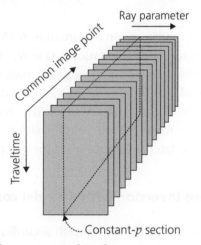

Figure 10.6 CRP gathers, presented in the ray-parameter (p) domain, are sorted into constant-p sections.

well-logging according to the subsurface structure. Based on Equation 10.14, we may define a set of equations as the following:

$$\xi_k(p) \equiv \frac{1}{2}\ln\frac{RI_k(p)}{RI_1(p)} = \sum_{i=1}^{k} r_i(p),$$
(10.21)

where $RI_k(p)$ is the ray impedance along a path with ray parameter p, defined by Equation 10.19. If presenting Equation 10.21 in the matrix-vector form as

$$\xi = Jr \,, \tag{10.22}$$

where the matrix J is a simple integral operator, and $r = \{r_k(p)\}$ is the vector of reflectivity series, then the objective function for the reflectivity inversion in the ray-parameter domain is

$$\phi(r) = (\tilde{s} - Gr)^T C_d^{-1}(\tilde{s} - Gr) + \mu_1 R(r,\lambda) + \mu_2(\xi - Jr)^T C_r^{-1}(\xi - Jr) \,, \tag{10.23}$$

where $R(r,\lambda)$ is the Cauchy model constraint, and C_r is the model covariance matrix. The least-squares solution of the reflectivity series is given by

$$r = \left(G^T C_d^{-1} G + \frac{\mu_1}{\lambda^2} D + \mu_2 J^T C_r^{-1} J \right)^{-1} (G^T C_d^{-1} \tilde{s} + \mu_2 J^T C_r^{-1} \xi) \,, \tag{10.24}$$

where D is a diagonal matrix, given by Equation 6.45.

In the single channel implementation, $G = W$, the wavelet matrix straightforwardly. In the multichannel implementation, matrix $G = WL$, where L is the spatial prediction operator, and vectors $\tilde{s}$ and r are replaced with matrices $\tilde{S}$ and R. Meanwhile, when replicating the constraint ξ to make a matrix, each column should has a weight, pre-assigned according to the lateral distance from the given reference position.

10.5 Ray-impedance inversion with a model constraint

Once seismic reflectivity traces are sorted according to ray-parameter values, ray-impedance inversion may be performed on constant ray-parameter sections, just like a conventional acoustic impedance inversion on a stack section. The objective function for the ray-impedance inversion is given as

$$\phi(z) = [r - f(z)]^T C_r^{-1}[r - f(z)] + \mu(z - z_0)^T C_z^{-1}(z - z_0) \,, \tag{10.25}$$

where r is the 'data' vector, consisting of reflectivity series obtained from a sparse inversion in the previous section, z is the impedance vector to be inferred, $\mathbf{z}_0$ is a reference model, $\mathbf{C}_r$ now is the data covariance matrix, and $\mathbf{C}_z$ is the model covariance matrix. The reflectivity synthetic $f(\mathbf{z})$ may be expressed as

$$r_k = \frac{z_{k+1} - z_k}{z_{k+1} + z_k}, \tag{10.26}$$

for $k = 1, 2, \cdots, N$. Setting $\partial\phi(\mathbf{z})/\partial\mathbf{z} = \mathbf{0}$, we get the model update as

$$\Delta\mathbf{z} = [\mathbf{F}^T\mathbf{C}_r^{-1}\mathbf{F} + \mu\mathbf{C}_z^{-1}]^{-1}\mathbf{F}^T\mathbf{C}_z^{-1}[\mathbf{r} - f(\mathbf{z})], \tag{10.27}$$

where $\mathbf{F}$ is the matrix of Fréchet derivatives of the reflectivity synthetic $f(\mathbf{z})$, with respect to the impedance parameters $\mathbf{z}$.

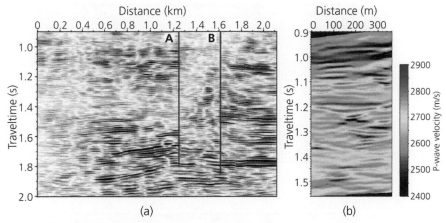

Figure 10.7 (*a*) A seismic section across two boreholes A and B. (*b*) Crosshole seismic velocity image obtained from waveform tomography.

Figure 10.7 displays an example seismic section across two vertical boreholes. The velocity image, obtained from crosshole seismic waveform tomography, may be used as a constraint in the following ray impedance inversion. The bulk density is calculated empirically (Gardner *et al.*, 1974).

Figure 10.8 displays a constant ray-parameter section ($p = 150$ ms/km), inverted reflectivity section, and the inversion result of ray impedance, using the crosshole velocity image as a model constraint.

It should be emphasised that geological models play essential roles in seismic inversion, both in the reflectivity inversion of Equation 10.23 and the impedance inversion of Equation 10.25.

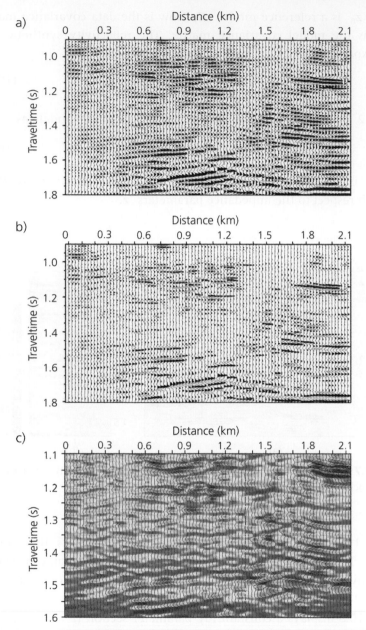

Figure 10.8 (*a*) A constant ray-parameter section (with p = 150 ms/km). (*b*) Inverted reflectivity section, corresponding to the constant p value. (*c*) The result of ray-impedance inversion, in which a crosshole seismic constraint is used.

CHAPTER 11

Seismic tomography based on ray theory

Seismic tomography is a group of seismic inversion methods based on data prediction via solution of a wave equation. This field developed rapidly in the 1980s in the form of traveltime inversion using seismic ray theory (Nolet, 1987). It is now undergoing another phase of rapid development as available computing power begins to make waveform tomography practicable. We shall cover waveform tomography in the following chapters. In the current chapter, we discuss seismic tomography based on ray theory, which essentially is a high frequency approximation to the wave equation.

Traveltime inversion has been used widely in regional and global geophysics (Rawlinson and Sambridge, 2003; Rawlinson et al., 2010; Zelt, 2011; Rawlinson et al., 2014). However, in exploration geophysics particularly with reflection geometry, traveltime inversion exists a well-known issue of ambiguity between velocity and reflector depth in the presence of velocity variations. Seismic inversion using only reflection traveltime data may not be possible to resolve this ambiguity. The inclusion of amplitude data in the inversion may help to resolve the ambiguity. We shall advocate the use of amplitude data, simultaneously with the traveltime data, in reflection seismic tomography.

11.1 Seismic tomography

The concept of tomography is based on the idea that an observed dataset consists of integrals along lines or rays of certain physical quantities. The

Seismic Inversion: Theory and Applications, First Edition. Yanghua Wang.
© 2017 John Wiley & Sons, Ltd. Published 2017 by John Wiley & Sons, Ltd.

purpose of tomography is to reconstruct a model of these physical quantities, such that the predicted data agree approximately with the measurements. Much research on the application of seismic tomography has been undertaken in the reflection configuration, as this has immediate relevance to much exploration work.

To imitate tomographic inversion closely, that is, line integration along rays, we could assume that the reflector is completely defined by *a priori* information, and use tomography only to infer the velocity field. Seismic depth migration would have been a good tool to elicit the reflector structure, before using traveltime inversion for imaging the velocity variation. An alternative approach is to invert for the velocity distribution simultaneously with reflector positions. The unknown reflectors are parameterised in some suitable fashion and are included in the inversion (Bishop *et al.*, 1985; Farra and Madariaga, 1988; Wang and Houseman, 1995). Since the extra parameters are not really tomographic, i.e. they are not line integrated in the same way as slowness (reciprocal velocity) is for traveltimes, we shall adopt a general optimisation approach if we wish to keep the two kinds of variables on a similar footing.

For an optimisation problem, the objective function can be defined by the data misfit:

$$\phi(\mathbf{m}) = [\mathbf{d} - G(\mathbf{m})]^{\mathrm{T}} \mathbf{C}_{\mathrm{d}}^{-1} [\mathbf{d} - G(\mathbf{m})] \,, \tag{11.1}$$

where $\mathbf{d}$ is the vector of data observation, $\mathbf{C}_{\mathrm{d}}$ is the data covariance matrix, $G(\mathbf{m})$ the forward prediction, and $\mathbf{m}$ is the vector of unknown model parameters that describe the properties of the subsurface media.

In this chapter, the 'data' are the traveltime and amplitude samples of the reflection seismic response. The inverse problem is very nonlinear, because small changes in the interface position produce large changes in ray trajectories. Thus, all source-receiver rays have to be retraced at each iteration of the inverse problem. For solving this nonlinear problem, a linearisation procedure includes a quadratic approximation to the objective function. Then, minimising this quadratic form will lead to a linear system.

11.2 Velocity-depth ambiguity in reflection tomography

In this chapter, where we focus on the ray theory-based inverse problem in the reflection configuration, we consider a stratified structure, consisting of continuously varying interfaces separating heterogeneous velocity layers.

When solving non-planar reflection surfaces and spatially variable velocity variations, there may be an ambiguity in the inversion solution in the form of a trade-off between reflector depth and velocity anomaly. Consider a constant velocity model above a flat horizontal reflector, except for a small area above the reflector with higher velocity than the surrounding region. If we assume that the velocity above the reflector is constant and the reflector is not necessarily flat, we might obtain a computed model whose traveltimes fit the measured data reasonably well but not exactly. The reflector of the computed model will be shallower under the velocity anomaly, a feature that may lead a structural geologist to suspect a trap for hydrocarbons. Thus inappropriate modelling can lead to incorrect conclusions concerning the model features that are of interest.

The ambiguity between the velocity and the depth is not caused by the particular inversion algorithm being used in the tomography but is a feature of the acquisition geometry and the geometry of the subsurface. Thus it is a pervasive problem in reflection seismology. The factors that control how an anomaly affects traveltimes (whether the anomaly correlates or anti-correlates with traveltimes or causes ambiguities) are its wavelength, its thickness and its height above a reflector. Other factors, such as the source-receiver offset and velocity anomaly magnitude, affect the magnitude of the traveltimes, but not their behaviour. If refractions are available, the ambiguity in reflection inversion can be reduced slightly because of the constraint of refraction travel paths.

The wavelength of the ambiguity changes with source-receiver offset. However, it is such a weak function of source-receiver separation that multiple offset processing, in practice, does little to resolve the ambiguity (Kosloff and Sudman, 2002). In this section, we shall analyse the ambiguity in traveltime inversion, and compare it with the ambiguity in amplitude inversion. We will see that, if using both traveltime and amplitude data in the inversion, the ambiguity between velocity variation and reflector depth can be reduced.

Consider a model with a planar reflecting interface at depth z and slowness u. A reflection for the source-receiver offset y is recorded at the surface. We wish to recover z and u from the data.

In traveltime inversion, the reflection traveltime equation is given by

$$T(u, z; y) = u\sqrt{y^2 + (2z)^2} \, . \tag{11.2}$$

The fractional variation in traveltime is then given in terms of the fractional changes in slowness and reflector depth,

$$\frac{\Delta T}{T} = \frac{\Delta u}{u} + \frac{\Delta z}{z}\cos^2\theta,$$ (11.3)

where θ is the angle of incidence. When θ is zero, that is, the offset is zero, there is no change in traveltime whenever the fractional depth change equals the fractional velocity change. The model is completely ambiguous. Therefore, $\cos^2\theta$, the angle of incidence, plays an important role in the reconstruction of the velocity and depth functions, and controls the degree of ambiguity between velocity variation and reflector depth.

In the case of one planar reflector model, a ray amplitude can be estimated as

$$A(u, z; y) \propto \frac{C}{D},$$ (11.4)

where C is the reflection amplitude coefficient and D is the geometrical spreading. The reflection amplitude coefficient can be approximated by (Wang and Houseman, 1995)

$$C(u, u_b; y) = \frac{u - u_b}{u + u_b}\eta(y),$$ (11.5)

where u_b is the wave slowness in the underlying layer and $\eta(y)$ is a factor relating to the incident angle, $0.5 \le \eta \le 1$. The geometrical spreading function is

$$D(z; y) = \sqrt{y^2 + (2z)^2}.$$ (11.6)

An evaluation of the fractional change in amplitude can be obtained as

$$\frac{\Delta A}{A} = \frac{1}{2C_0}\frac{\Delta u}{u} - \frac{\Delta z}{z}\cos^2\theta,$$ (11.7)

where $\Delta A/A = \Delta\ln A$ is the perturbation of logarithmic amplitude, and $C_0 \approx (u - u_b)/(u + u_b)$ is the reflection amplitude coefficient at zero incident angle. Equation 11.7 suggests that the amplitude inversion also shares, with the traveltime inversion, the problem of ambiguity between velocity variation and reflector depths, also characterised by $\cos^2\theta$.

Regarding the ambiguity issue, Equation 11.3 can be taken as a characteristic equation in traveltime inversion, and Equation 11.7 can be taken as a characteristic equation in amplitude inversion. Using either traveltimes or amplitudes alone cannot solve the problem. However, if we simply make an algebraic addition of Equation 11.3 and Equation 11.7, the ambiguity term $\cos^2\theta$ can be cancelled. This is a joint inversion.

To understand this situation in physical terms, let us consider again the model with constant velocity above a flat horizontal reflector. If the reflector were slightly shallower, the velocity above it would be slightly smaller so that the reflection traveltimes in this perturbed model would almost equal those of the original model, but if we want perturbed amplitudes equal to those of the original model, the velocity would be slightly greater. Constrained by both types of data, a compromise velocity would be produced, subject to the forms adopted for the data variations and model parameterisation.

This joint inversion method is an intermediate step between traveltime inversion and waveform inversion. As a practical application, it uses more information than traveltime inversion to model subsurface structure, and will provide better velocity resolution than is possible with traveltime data alone, without excessive use of computational resources.

11.3 Ray tracing by a path bending method

In a ray theory-based tomographic inversion it is necessary to have a robust ray tracing routine.

For traveltime and amplitude inversion, one group of forward calculation methods is grid-based schemes, which involve the calculation of traveltimes at all regular grids within the velocity model. They are often based on the eikonal equation (Vidale, 1990; van Trier and Symes, 1991; Qin *et al.*, 1992; Hole and Zelt, 1995; Kim, 2002). Wavefronts and rays can be obtained *a posteriori* for computing quantities other than traveltime between any two points. Wang (2013) proposed a system of equations for computing simultaneously the traveltime field and the slowness paths normal to wavefronts in anisotropic media. However, grid based schemes only compute the first-arrival traveltimes in most cases.

A popular group of methods is seismic ray tracing, which determines trajectories along which seismic energy propagates between sources and receivers (Julian and Gubbins, 1977; Červený, 2001; Červený *et al.*, 2007;

Rawlinson *et al.*, 2007). This numerically stable method in turn leads to an efficient and accurate computation of traveltimes and amplitudes along the energetic ray paths. In this section, we present a path bending method for the two-point ray tracing within a heterogeneous media. This path bending method, based on Fermat's principle, is reduced to a solution of a linearised tridiagonal equation system.

For the calculation of ray amplitudes, the model velocity distribution must vary smoothly within a layer, in which any ray path is composed of an arc with continuously varying curvature. The traveltime and its derivatives with respect to the model parameters can be calculated and, if the ray-tube around the reference ray smoothly diverges, calculation of the ray-amplitude is also stable. At an interface between layers, the assumption of a smooth interface (i.e. the existence of its partial derivatives of first order and second order) is necessary in order to calculate the transformation of the ray and the paraxial rays.

According to Fermat's principle, the ray path is the path γ which minimises the traveltime T, given by

$$T(\gamma) = \int_{\gamma} \frac{d\ell}{v} \to \min, \tag{11.8}$$

where v is velocity and ℓ is the ray arc-length. For computational purposes, the ray path under consideration can be discretised into a polygonal path,

$$\gamma = \{X_0, X_1, \cdots, X_K, \cdots, X_{2K}\} \tag{11.9}$$

consisting of 2K+1 points in three-dimensional space, numbered from 0 to 2K, and connected by straight line segments or circular arcs. The traveltime can be expressed explicitly as

$$T = \frac{1}{2} \sum_{i=1}^{2K} (u_{i-1} + u_i) \Delta \ell_i, \tag{11.10}$$

where $u_i = 1/v_i$ and $\Delta \ell_i$ is the length of the ray segment between points X_{i-1} and X_i. For ray tracing, we may consider X_K as the reflection point, and fix the end points, X_0 and X_{2K}, of the ray path. Fermat's principle states that seismic energy travels along a path for which the first order variation with all neighbouring paths is zero (Sheriff, 1991):

$$\nabla_\gamma(\gamma) = 0. \tag{11.11}$$

Consider a layered structure with M interfaces defined by

$$f_k(X) = 0, \qquad \text{for} \quad k = 1, \cdots, M. \tag{11.12}$$

We divide each layer into N_k thin layers and insert N_{k-1} interpolation levels between each pair of interfaces, $f_k(X) = 0$ and $f_{k+1}(X) = 0$, by performing a linear interpolation in the vertical direction:

$$z_j = z_k + \frac{j}{N_k}(z_{k+1} - z_k), \qquad \text{for} \quad j = 1, \cdots, N_{k-1}. \tag{11.13}$$

The jth interpolation level (Figure 11.1a) is then specified by a smooth surface defined by means of cubic spline interpolation among a set of discrete depth values $z_j(x,y)$.

Assume a reflected ray trajectory intersects a total of K of interfaces and interpolation levels, and the intersection points are ordered as $1, 2, \cdots,$ $2K - 1$. The free parameters are $\xi = \{\xi_1, \cdots, \xi_K, \cdots, \xi_{2K-1}\}$, where ξ includes x or y components of the intersection points. Thus, the number of free parameters is $2(2K - 1)$. The corresponding gradient of traveltime $\nabla_\xi T(\gamma)$ can be written explicitly as

$$\frac{\partial T}{\partial \xi_i} = \frac{1}{2}\left[(\Delta\ell_i + \Delta\ell_{i+1})\left(\frac{\partial u_i}{\partial \xi_i} + \frac{\partial u_i}{\partial z_i}\frac{\partial z_i}{\partial \xi_i}\right) + \frac{u_i + u_{i+1}}{\Delta\ell_{i+1}}\left((\xi_i - \xi_{i+1}) + (z_i - z_{i+1})\frac{\partial z_i}{\partial \xi_i}\right) \right.$$
$$\left. + \frac{u_i + u_{i-1}}{\Delta\ell_{i-1}}\left((\xi_i - \xi_{i-1}) + (z_i - z_{i-1})\frac{\partial z_i}{\partial \xi_i}\right)\right], \tag{11.14}$$

for $i = 1, 2, \cdots, 2K - 1$. Equation 11.14 gives a tridiagonal linear equation system in the set of unknowns $\{\xi_i\}$, which can be solved iteratively.

However, seismic ray tracing is a nonlinear system, because any path perturbation will cause velocity changes and these velocity changes will further affect the ray path. Wang (2014) proposed to enforce Fermat's minimum-time principle as a constraint for the solution update, instead of conventional error minimisation in the nonlinear system. As the algebraic problem is incorporated with the physical principle, the solution for seismic ray tracing is stabilised.

As an example, Figure 11.1b shows the ray paths through a 2-D example with arbitrary slowness and interface variation. This algorithm is robust for rays whose angle of inclination relative to the interpolation surfaces is not too shallow. Because the ray path is approximated by straight-line segments, the algorithm should not be expected to give accurate results in the vicinity of turning rays.

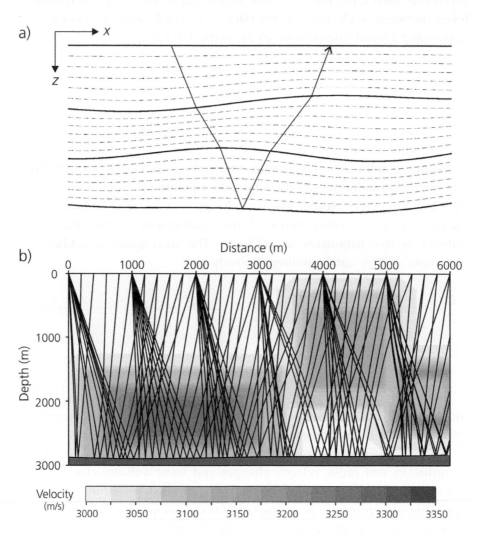

Figure 11.1   (*a*) A layered Earth structure with interpolation between interfaces. A reflected ray trajectory intersects all the interfaces and interpolated levels. (*b*) The geometry of seismic reflection rays within a variable velocity medium, reflected from a curved interface.

11.4 Geometrical spreading of curved interfaces

The structural amplitude effect is included in the geometrical spreading of rays. The pattern of 'spherical divergence' depends on the way that the velocity distribution varies with depth and is also affected by any interface. There are two stages of such spreading, the first from the source to the reflector and the second from the reflector (which can be viewed as a virtual source) to the receiver (Figure 11.2). The character of the observed reflections depends strongly on the shape of the reflector. When the curvature of the reflector is greater than that of the incident wavefront, we have the possibility of the formation of subsurface focusing, with possible phase distortion of the observed reflection.

The amplitude variation due to geometrical spreading is given by

$$A(\ell) = A_0 C \frac{\ell_0}{L(\ell)}, \tag{11.15}$$

where $L(\ell)$ is the ray geometrical spreading function, ℓ is the distance from a source point measured along the ray path, C is related to the changes due to the acoustic impedance contrasts across interfaces using the Zoeppritz relations, and A_0 is taken to be the amplitude of the wave at some distance ℓ_0 sufficiently close to the source that there are no intervening interfaces, but sufficiently far that near-source effects can be neglected and the wavefront is spherical.

Let ℓ_i represent the length of the ray segment between two points N_{i-1} and N_i, where N_0 and N_{K+1} represent the source and the receiver points. Because of the interface curvature and the impedance contrast, the distance between the incidence point N_i and the virtual image of the source N_0' is ℓ_i', whereas the provisional observation position is at N_{i+1}'. In the perpendicular plane the interface curvature is zero because of the assumption of 2-D structure and the distance between N_i and N_0' is ℓ_i''. The geometrical spreading function can be expressed as follows (Wang and Houseman, 1994):

$$L(\ell) = \ell_1 \prod_{i=1}^{K} \left[\left(1 + \frac{\ell_{i+1}}{\ell_i'} \right) \left(1 + \frac{\ell_{i+1}}{\ell_i''} \right) \right]^{1/2},$$

$$\text{for } \ell_{K+1} = \ell - \sum_{i-1}^{K} \ell_i, \tag{11.16}$$

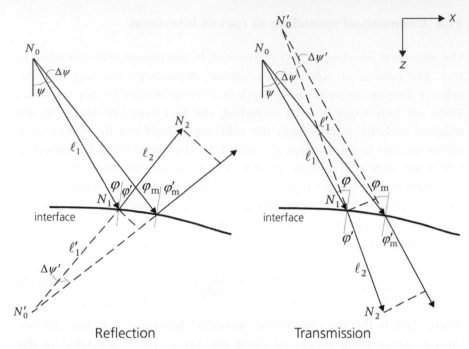

Reflection   Transmission

Figure 11.2 Geometry of incidence and reflection (or refraction), φ and φ' are the angles of incidence and reflection (or refraction) for a ray of take-off angle ψ, φ_{m} and φ'_{m} represent modified angles where the ray take-off angle is $\psi + \Delta\psi$. N_0 represents the source point, N'_0 is the virtual image of the source, N_1 is the incident point, and N_2 is the initial observation point. The distance between N_0 and N_1 is ℓ_1, between N'_0 and N_1 is ℓ'_1, and between N_1 and N_2 is ℓ_2.

where K is the number of intersection points of a specified ray with successive interfaces.

The virtual ray distance ℓ'_i in the ray plane is given by

$$\frac{1}{\ell'_i} = \frac{1}{(\ell'_{i-1} + \ell_i)} \frac{v_{i+1}}{v_i} \frac{\cos^2 \varphi_i}{\cos^2 \varphi'_i} + \frac{1}{\cos \varphi'_i} \left(\frac{v_{i+1}}{v_i} \frac{\cos \varphi_i}{\cos \varphi'_i} \pm 1 \right) \Theta_i, \qquad (11.17)$$

where the '+' sign refers to the reflection case and the '−' sign refers to the refraction case, v_i is the local constant velocity along ray segment ℓ_i, φ_i and φ'_i are incident and reflection or refraction angles at the point N_i, and Θ_i is the factor describing the effect of local curvature of the ith interface, defined by

$$\Theta_i(x) = \pm \frac{d^2 z_i}{dx^2} \left[1 + \left(\frac{dz_i}{dx} \right)^2 \right]^{-3/2}, \qquad (11.18)$$

where here the '$\pm$' signs refer to the incident ray with an acute or an obtuse angle with the z axis (or to the down- and up-going rays, respectively). The interface is represented by a single-valued function $z_i(x)$ in which x is the horizontal coordinate and z is depth below some reference level at coordinate x.

In the perpendicular direction, Equation 11.17 with $\Theta_i = 0$ and $\varphi_i = 0$ can be used to define the apparent distance ℓ_i'' to the virtual image of the source:

$$\frac{1}{\ell_i''} = \frac{v_{i+1}}{\displaystyle\sum_{j=1}^{i} \ell_j v_j}. \qquad (11.19)$$

The explicit analytical expression above clearly indicates that reflection seismic amplitude data do contain information on reflector geometry. The tomographic inversion using amplitude data will, in particular, extract the information on reflector geometry that causes the focusing and defocusing of the ray amplitude.

11.5 Joint inversion of traveltime and amplitude data

In the inversion, let us consider a stratified structure model with smooth and curved interfaces separating homogeneous layers. Within each layer the *P*-wave velocity is assumed to be constant. However, the inversion allows the elastic properties to vary along the horizontal interfaces (Figure 11.3). This model parameterisation is based on the following assumptions:

1) Amplitude data do not constrain absolute interval velocities, which are essentially constrained by reflection traveltimes.

2) The stack of constant velocity layers is a good macro-model in a specific area as far as the reflection traveltimes are concerned.

3) Amplitudes are significantly sensitive to velocity perturbations in the vicinity of a reflection point, when compared with the velocity variation along the whole ray path from source to receiver.

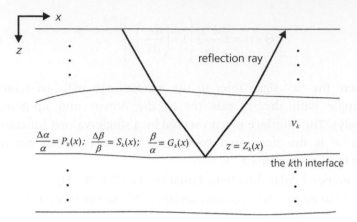

Figure 11.3 A 2-D stratified structure model considered in the joint inversion of traveltime and amplitude data, inverting simultaneously for the interface geometry and the elastic parameters along the interfaces.

The reflection and transmission coefficients can be expressed as the following (Wang, 1999a):

$$R_{PP} \approx \frac{1}{2}\left[r+1+\tan^2\theta - \left(\frac{\beta}{\alpha}\right)^2 \sin^2\theta \right]\frac{\Delta\alpha}{\alpha} - 4\left(\frac{\beta}{\alpha}\right)^2 \sin^2\theta \frac{\Delta\beta}{\beta}$$

$$+\left(\frac{\beta}{\alpha}\right)^3 \cos\theta\sin^2\theta\left(r\frac{\Delta\alpha}{\alpha}+2\frac{\Delta\beta}{\beta}\right)^2, \tag{11.20}$$

and

$$T_{PP} \approx 1 - \frac{1}{2}(r+1-\tan^2\theta)\frac{\Delta\alpha}{\alpha} - \left(\frac{\beta}{\alpha}\right)^3 \cos\theta\sin^2\theta\left(r\frac{\Delta\alpha}{\alpha}+2\frac{\Delta\beta}{\beta}\right)^2, \tag{11.21}$$

where $\Delta\alpha/\alpha$ is the relative P-wave velocity contrast, $\Delta\beta/\beta$ is the relative S-wave velocity contrast, β/α is the ratio of average S- to P-wave velocities, and r is the empirical relationship between the density contrast $\Delta\rho/\rho$ and the P-velocity contrast:

$$r = \frac{\Delta\rho/\rho}{\Delta\alpha/\alpha}. \tag{11.22}$$

This r value can be 0.261, 0.265, 0.225, 0.143, 0.160 for sand, shale, limestone, dolomite and anhydrite, respectively (Wang, 2003b).

Thus, the elastic property of a reflector is measured in terms of three elastic parameters, laterally variable along the interface, denoted by

$$\frac{\Delta\alpha}{\alpha} = P(x), \qquad \frac{\Delta\beta}{\beta} = S(x), \qquad \frac{\beta}{\alpha} = G(x), \qquad (11.23)$$

where x is the spatial coordinate. In a simultaneous inversion for both the interface geometry and the elastic parameters, both the geometrical spreading for a multi-layered structure and the reflection/transmission coefficients at interfaces are represented analytically as function of the interface geometry, $z = Z(x)$, and the elastic parameters, $P(x)$, $S(x)$ and $G(x)$, along a target reflector.

The depth of an interface and the variation of the elastic parameters along the horizon, $\{ Z(x), P(x), S(x), G(x) \}$, are parameterised by a truncated Fourier series,

$$f(x) = a_0 + \sum_{n=1}^{N} \left[a_n \cos(n\pi k_0 x) + b_n \sin(n\pi k_0 x) \right], \qquad (11.24)$$

where k_0 is the basis wavenumber and $\{a_0, a_n, b_n, (n = 1, N)\}$ are amplitude coefficients of the harmonic terms. The model vector, $\mathbf{m}$, that we invert for in a nonlinear inversion thus consists of four sets of amplitude coefficients

$$\{a_0^{(J)}, a_n^{(J)}, b_n^{(J)}, \quad \text{for } n = 1, \cdots, N\}, \qquad (11.25)$$

where $J = 1, \cdots, 4$ representing Z, P, S and G, respectively. In the following inversion tests, we set the number of harmonic terms to $N = 20$. Including the interval velocity above each reflector, there are $1 + 4(2N + 1)$ parameters in total in the inversion. The data vector, $\mathbf{d}$, in the inversion consists of reflection traveltimes and reflection amplitudes (Wang, 1999c; Wang and Pratt, 2000).

The simultaneous inversion for model geometry and elastic parameters is demonstrated using real seismic reflection data from the North Sea. Figure 11.4 shows an example profile, after prestack time migration, and indicates five reflections R_i ($i = 1, 2, \cdots, 5$). Close examination of Figure 11.4 clearly shows that there are gradual phase changes at points A and B, on reflection 3 and 4, respectively, and phase changes at C and D in reflection 5, where the reflection event varies from single wavelet to multi-wavelet. These phase changes could correspond to changes in the elastic properties along the reflections.

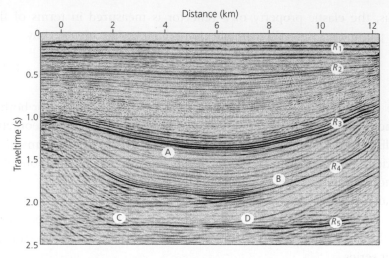

Figure 11.4 Reflection seismic profile from the North Sea. Traveltimes and amplitudes, extracted from migrated CRP gathers, are used in the inversion for interface geometry and elastic parameters along reflectors. Four reflections $R_2, \cdots,$ R_5 are considered in the inversion. Labels A, B, C and D indicate phase changes of reflection events.

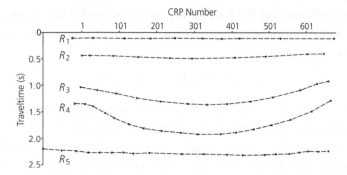

Figure 11.5 Reflection times picked from the reference profile in Figure 11.3. The reflection time is used as a reference time in a cross-correlation procedure for picking the relative reflection traveltimes and amplitudes on individual traces of CRP gathers.

From this stack profile we pick the reflection times, which in turn are used as reference traveltimes in a cross-correlation process for extracting traveltimes and amplitudes from traces in any time-migrated common-reflection-point (CRP) gather. For this example dataset, Figure 11.5 plots the reference traveltimes of five reflections (including the sea-bottom reflector). There are a total of 601 CRPs considered.

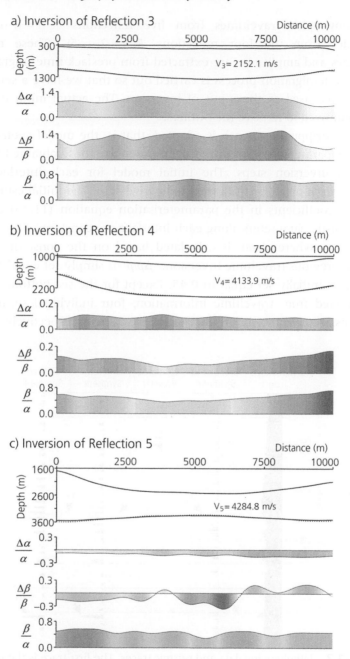

Figure 11.6 Simultaneous inversion of reflections 3, 4 and 5. In each case, the inversion model is consisted of a constant interval velocity, the interface geometry z and three elastic parameters ($\Delta\alpha/\alpha$, $\Delta\beta/\beta$, β/α). The interface geometry (solid line) is compared with inversion result from traveltime inversion (dotted line).

We now pick traveltimes from five reflections R_i $(i = 1, \cdots, 5)$ and amplitudes from the four reflections R_i $(i = 2, \cdots, 5)$. These reflection traveltimes and amplitudes are extracted from prestack time-migrated CRP gathers. A demigration process is carried out so that we have a set of 'true' observations, used as the input of inversion. The actual input data have been winnowed to remove the outliers.

A layer-stripping approach is adopted, that is, the model geometry and the elastic parameters used in the overburden are those obtained from the preceding inversion steps. The initial model for each interface is an interface curve obtained from traveltime inversion. The initial estimates for the three coefficients in the parameterisation equation (11.25) are zero-valued, elastic parameters along each interface are constants (i.e. all except that of a_0), where $\Delta\alpha/\alpha$ is calculated based on the constant velocities from the previous traveltime inversion, $\Delta\beta/\beta$ is simply set equal to $\Delta\alpha/\alpha$, and β/α is initially set equal to 0.45. Except for the sea bottom, which is reconstructed from traveltime information, four individual simultaneous inversions are performed, for reflections 2, 3, 4 and 5, respectively.

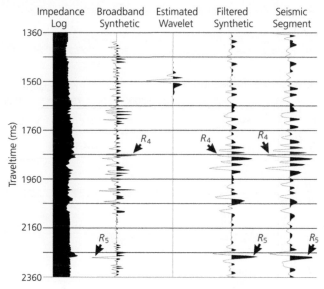

Figure 11.7 Lithology log data and seismic traces. The first trace is the impedance log. The second trace is the broadband synthetic computed from well logs. The third trace is the estimated seismic wavelet. The fourth trace is the filtered synthetic, filtered by estimated seismic wavelet. The fifth trace is the field seismic segment. The synthetic and field seismic traces match accurately.

Figure 11.6 dipects the inversion results for reflections 3, 4 and 5. In each of these cases, an inversion model is consisted of a constant interval velocity, the interface geometry and three elastic parameters $\Delta\alpha/\alpha$, $\Delta\beta/\beta$, and β/α. The inverted interface geometry (solid line) is compared with the traveltime inversion result (dotted line). The horizontal variations in the elastic parameters are shown by both curves and the grey scale, which facilitates comparison of their spatial variations.

At reflector R_4, $\Delta\beta/\beta$ and β/α are correlated. But at reflectors R_3 and R_5, they do not show strong correlation. The correlation could be partly due to intrinsic variation of elastic properties, although any solution artefact is possible.

Lithology log data (Figure 11.7) revealed that reflector R_4 is an interface between muddy chalk and chalk. Normally, $\Delta\beta > 0$, and β/α is expected to be around 0.5. Reflector R_5 is the interface between evaporites (dolomite/salt) and shale. Thus $\Delta\beta < 0$ and β/α would be about 0.55–0.60. The results shown in Figure 11.6 appear to be quite close to these expectations.

CHAPTER 12

Waveform tomography for the velocity model

Waveform tomography is a seismic inversion method that extracts sliced images of the Earth's subsurface from observed seismic waveforms. It is also referred to as full waveform inversion, and is increasingly done in three dimensional these days (Warner *et al.*, 2013).

Seismic waveform tomography has been used successfully on transmission data, such as crosshole seismic data (Pratt and Worthington, 1990; Pratt, 1999; Wang and Rao, 2006; Rao *et al.*, 2016). It also has been applied to wide-angle reflection/refraction seismic data with a certain degree of success (Bunks *et al.*, 1995; Pratt *et al.*, 1996; Ravaut *et al.*, 2004; Operto *et al.*, 2006; Bleibinhaus *et al.*, 2007; Brenders and Pratt, 2007; Morgan *et al.*, 2013). In addition, waveform tomography has been used for regional-scale studies on the crustal and upper mantle velocity structure, using scattering waves, surface waves and SH-waves from either exploration seismics or broad-band teleseismograms (Helmberger *et al.*, 2001; Pollitz and Fletcher, 2005; Priestley *et al.*, 2006; Fichtner, 2010).

This chapter introduces seismic waveform tomography in the reflection acquisition geometry, which records reflection seismic data at the Earth's surface with limited source-receiver offsets. Reflection seismic is a routine method in hydrocarbon exploration, as the data dominated by the pre-critical reflection energy, reflected back from subsurface contrasts in physical parameters, are well suited for seismic migration for the structural image. Therefore, it would be desirable if these seismic reflection data can be used to quantitatively extract the physical parameters of the Earth's interior. These parameters are essential for identifying different lithologies,

Seismic Inversion: Theory and Applications, First Edition. Yanghua Wang.
© 2017 John Wiley & Sons, Ltd. Published 2017 by John Wiley & Sons, Ltd.

different fracture characteristics, and even for indicating the hydrocarbon distribution directly.

However, when applying waveform tomography to limited-offset reflection seismic data, it is difficult to recover long-wavelength variations in velocity, since reflection waveforms are highly sensitive to the contrasts in velocity, and are not so sensitive to long-wavelength variations (Mora, 1987, 1988). It is possible to decouple the smooth reference velocity and the short-wavelength variations by performing a waveform-fit in an alternating fashion (Snieder *et al.*, 1989). It is also feasible to invert first for the short-wavelength variation in velocity and then for long-wavelength variation with a suitable re-parameterisation to much fewer model parameters. In any case, it is common to use traveltime and/or amplitude inversion to recover the long-wavelength velocity variation, and this serves as a good starting model for the iterative waveform inversion.

Another difficulty, when using reflection seismic data in waveform tomography, is to recover the velocity variation in the deep part of the model. There are a number of causes. First, seismic reflection amplitudes decrease with increasing depth, and near-surface waveforms dominate the data-fitting in the inversion. In other words, near-surface velocity updates have much more influence on the data fitting than deeper changes in velocity. Second, ray-density in the deep part is much lower than that in the shallow part of the model. Tomography is an integral along ray paths, on which data residuals are evenly distributed. Therefore, high ray-density leads to a large model update, and low ray-density has a small velocity update in the iterative solution. To tackle the problem, we shall introduce a practical weighting scheme that scales velocity updates according to the depth (Wang and Rao, 2009).

12.1 Inversion theory for waveform tomography

The objective function for this inverse problem is defined by data misfit as

$$\phi(\mathbf{m}) = \frac{1}{2}[\mathbf{u}_{obs} - \mathbf{u}(\mathbf{m})]^{H}\mathbf{C}_{d}^{-1}[\mathbf{u}_{obs} - \mathbf{u}(\mathbf{m})], \qquad (12.1)$$

where $\mathbf{m}$ is the model to invert for, $\mathbf{u}(\mathbf{m})$ is a modelled data set, $\mathbf{u}_{obs}$ is an observed data set, the superscript ^{H} denotes the Hermitian transpose, the transpose of the complex conjugate, and $\mathbf{C}_{d}$ is the covariance operator in the data space with units of $(data)^2$, reflecting the uncertainties in the observed data set.

To minimise the objective function, a gradient-based method may be used. The gradient of the data misfit is defined by

$$\gamma \equiv \frac{\partial \phi}{\partial \mathbf{m}} = -\mathbf{F}^H \mathbf{C}_d^{-1} \Delta \mathbf{u} = -\mathbf{F}^H \Delta \hat{\mathbf{u}} , \qquad (12.2)$$

where $\mathbf{F}$ is the matrix of the Fréchet derivatives of the seismic waveform $\mathbf{u}(\mathbf{m})$, with respect to the model parameters, $\Delta \mathbf{u} = \mathbf{u}_{obs} - \mathbf{u}(\mathbf{m})$ is the data residual vector, and $\Delta \hat{\mathbf{u}} = \mathbf{C}_d^{-1} \Delta \mathbf{u}$ is a weighted data residual vector. The steepest ascent direction is $\hat{\gamma} = \mathbf{C}_m \gamma$, where $\mathbf{C}_m$ is the model covariance matrix with units of $(model\ parameter)^2$. The model update, following the steepest descent direction $-\hat{\gamma}$, is given by

$$\Delta \mathbf{m} = -\alpha \hat{\gamma} , \qquad (12.3)$$

where α is a step length that needs to be determined.

For evaluating the gradient γ using Equation 12.2, we need to know the Fréchet matrix $\mathbf{F}$, which relates the data perturbation $\Delta \mathbf{u}$ to the model perturbation $\Delta \mathbf{m}$ as

$$\Delta \mathbf{u} = \mathbf{F} \Delta \mathbf{m} . \qquad (12.4)$$

This is the first term in a Taylor's series of $\Delta \mathbf{u}$. However, the direct computation of $\mathbf{F}$ is a formidable task when $\mathbf{u}$ are seismic waveforms. Instead, the action of matrix $\mathbf{F}^H$ on the weighted data residual vector $\Delta \hat{\mathbf{u}}$ (Equation 12.2) can be computed by a series of forward modelling steps, as summarised in what follows.

The frequency-domain acoustic wave equation for a constant-density medium with velocity $v(\mathbf{r})$ is

$$\left(\nabla^2 + \frac{\omega^2}{v^2(\mathbf{r})} \right) u(\mathbf{r}) = -\delta(\mathbf{r} - \mathbf{r}_s) s(\omega) , \qquad (12.5)$$

where $\mathbf{r}$ is the position vector, $\mathbf{r}_s$ indicates a source position, $s(\omega)$ is the source signature of frequency ω, and $u(\mathbf{r})$ is the (pressure) wavefield of this frequency. Wave equation (12.5) can be discretised using a finite difference scheme and represented in a matrix-vector form as

$$\mathbf{A}\mathbf{u} = -\mathbf{s} , \qquad (12.6)$$

where $\mathbf{A}$ is a matrix consisted of the frequency and the model property $\mathbf{m} = \{v(\mathbf{r})\}$, the velocity field only in this case, $\mathbf{u}$ is a vector representing the wavefield at all grids, and $\mathbf{s}$ is the source vector, which is zero everywhere except at the source location $\mathbf{r}_s$.

Performing a first-order derivative to Equation 12.6, with respect to a parameter m_i, we have

$$\mathbf{A}\frac{\partial \mathbf{u}}{\partial m_i} + \frac{\partial \mathbf{A}}{\partial m_i}\mathbf{u} = 0, \tag{12.7}$$

where $i \equiv (i_x, i_z)$ in two-dimensional (2-D) grids. We obtain explicitly a column vector of the Fréchet matrix as

$$\frac{\partial \mathbf{u}}{\partial m_i} = -\mathbf{A}^{-1}\frac{\partial \mathbf{A}}{\partial m_i}\mathbf{u}, \tag{12.8}$$

and in turn an element of the gradient vector $\boldsymbol{\gamma}$ as

$$\frac{\partial \phi}{\partial m_i} = \mathbf{u}^{\mathrm{H}}\left(\frac{\partial \mathbf{A}}{\partial m_i}\right)^{\mathrm{H}}(\mathbf{A}^{-1})^{\mathrm{H}}\Delta\hat{\mathbf{u}}. \tag{12.9}$$

Matrix $\partial\mathbf{A}/\partial m_i$ has the same dimensional size as matrix $\mathbf{A}$, but it has only a single nonzero element at grid $i \equiv (i_x, i_z)$:

$$\left[\frac{\partial \mathbf{A}}{\partial m_i}\right]_{i,i} = -\frac{2\omega^2}{v_i^3}. \tag{12.10}$$

Then, we can present

$$\mathbf{u}_{\mathrm{b}} = -(\mathbf{A}^{-1})^{\mathrm{H}}\Delta\hat{\mathbf{u}}. \tag{12.11}$$

The conjugate represents time reversal, and the inverse matrix $\mathbf{A}^{-1}$ multiplying to the data residual vector means using data residual $\Delta\hat{\mathbf{u}}$ as a virtual source to generate wavefield, $\mathbf{u}_{\mathrm{b}}$. Thus, the operation of Equation 12.11 is referred to as data-residual back-propagation.

Therefore, the gradient element in Equation 12.9 becomes

$$\frac{\partial \phi}{\partial m_i} = \overline{\left(\frac{2\omega^2}{v_i^3}\right)}\bar{u}_i u_{\mathrm{b},i}, \tag{12.12}$$

where $\overline{2\omega^2/v_i^3}$ is the complex conjugate of $2\omega^2/v_i^3$, and $\bar{u}_i$ is the complex conjugate of u_i.

Considering that there are multiple sources, and taking a summation over sources (s), Equation 12.12 becomes

$$\gamma(\mathbf{r}) = \left(\frac{2\omega^2}{v^3(\mathbf{r})}\right) \sum_s \overline{\mathbf{u}}(\mathbf{r};\mathbf{r}_s)\mathbf{u}_b(\mathbf{r};\mathbf{r}_s). \tag{12.13}$$

In Equation 12.13, the multiplication of the conjugated forward wavefield $\overline{\mathbf{u}}(\mathbf{r};\mathbf{r}_s)$ and the back-propagation wavefield $\mathbf{u}_b(\mathbf{r};\mathbf{r}_s)$ corresponds to a zero-lag correlation in the time domain, separately for each source (Tarantola, 1984, 1987).

The back-propagation equation (12.11) can be rewritten as

$$\mathbf{A}\overline{\mathbf{u}}_b = -\overline{\Delta\hat{\mathbf{u}}}, \tag{12.14}$$

where $\overline{\Delta\hat{\mathbf{u}}}$ is the complex conjugate of virtual source $\Delta\hat{\mathbf{u}}(\mathbf{r};\mathbf{r}_s)$, and $\overline{\mathbf{u}}_b$ is the complex conjugate of back propagation wavefield $\mathbf{u}_b(\mathbf{r};\mathbf{r}_s)$. This is the same equation as Equation 12.5. That is, wavefield $\overline{\mathbf{u}}_b(\mathbf{r};\mathbf{r}_s)$ is computed using the same forward modelling scheme as Equation 12.5, and using $\overline{\Delta\hat{\mathbf{u}}}(\mathbf{r};\mathbf{r}_s)$ as the source.

In summary, the frequency-domain waveform tomography is performed iteratively. At each iteration, the inversion procedure may be divided into five steps:

1) For a given initial model, calculating the synthetic wavefield $\mathbf{u}(\mathbf{r};\mathbf{r}_s)$ and data residual $\Delta\mathbf{u}$;

2) Back-propagating the weighted data residual $\Delta\hat{\mathbf{u}} = \mathbf{C}_d^{-1}\Delta\mathbf{u}$ as a virtual source to generate wavefield $\mathbf{u}_b(\mathbf{r};\mathbf{r}_s)$;

3) Correlating the forward wavefield $\mathbf{u}(\mathbf{r};\mathbf{r}_s)$ and the back-propagation wavefield $\mathbf{u}_b(\mathbf{r};\mathbf{r}_s)$;

4) Repeating steps 1–3 for every source, and summarising the zero-lag correlations of all sources to form the gradient direction γ;

5) Estimating the model update $\Delta\mathbf{m} = -\alpha\mathbf{C}_m\gamma$, where α is the optimal step length that can be found in the following section.

12.2 The optimal step length

For this gradient-based inversion method, the step length α should be carefully determined for the model update. Let us rewrite the model update in Equation 12.3 to

$$\Delta \mathbf{m} = -\alpha \mathbf{a} , \qquad (12.15)$$

where $\mathbf{a}$ is a normalised vector:

$$\mathbf{a} = \frac{\hat{\boldsymbol{\gamma}}}{\| \hat{\boldsymbol{\gamma}} \|} = -\frac{\mathbf{C}_m \mathbf{F}^H \mathbf{C}_d^{-1} \Delta \mathbf{u}}{\| \mathbf{C}_m \mathbf{F}^H \mathbf{C}_d^{-1} \Delta \mathbf{u} \|} . \qquad (12.16)$$

For the objective function in Equation 12.1, we make a quadratic approximation, as

$$\phi(\mathbf{m} + \Delta \mathbf{m}) = \phi(\mathbf{m}) + \Delta \mathbf{m}^T \boldsymbol{\gamma} + \frac{1}{2} \Delta \mathbf{m}^T \mathbf{H} \Delta \mathbf{m} , \qquad (12.17)$$

where $\mathbf{H}$ is the Hessian matrix of the objective function, and can be defined by

$$\mathbf{H} = \frac{\partial^2 \phi}{\partial \mathbf{m}^2} = \frac{\partial \boldsymbol{\gamma}}{\partial \mathbf{m}} = \mathbf{F}^H \mathbf{C}_d^{-1} \mathbf{F} - \frac{\partial \mathbf{F}^H}{\partial \mathbf{m}} \mathbf{C}_d^{-1} \Delta \mathbf{u} . \qquad (12.18)$$

Substituting the model update (Equation 12.15) into the quadratic function in Equation 12.17, and minimising this quadratic function by setting $\partial \phi / \partial \alpha = 0$, we determine the step length α as

$$\alpha = \frac{\mathbf{a}^T \boldsymbol{\gamma}}{\mathbf{a}^T \mathbf{H} \mathbf{a}} . \qquad (12.19)$$

Therefore, the model update is

$$\Delta \mathbf{m} = -\frac{\mathbf{a}^T \mathbf{a}}{\mathbf{a}^T \mathbf{H} \mathbf{a}} \boldsymbol{\gamma} . \qquad (12.20)$$

The Hessian matrix can be constructed using BFGS (Broyden-Fletcher-Goldfarb-Shanno) updating formula:

$$\mathbf{H}^{(k+1)} = \mathbf{H}^{(k)} - \frac{\mathbf{H}^{(k)} \mathbf{z}_k \mathbf{z}_k^H \mathbf{H}^{(k)}}{\mathbf{z}_k^H \mathbf{H}^{(k)} \mathbf{z}_k} + \frac{\mathbf{y}_k \mathbf{y}_k^H}{\mathbf{y}_k^H \mathbf{z}_k} , \qquad (12.21)$$

where

$$\begin{aligned} \mathbf{y}_k &= \boldsymbol{\gamma}^{(k+1)} - \boldsymbol{\gamma}^{(k)} , \\ \mathbf{z}_k &= \mathbf{m}^{(k+1)} - \mathbf{m}^{(k)} . \end{aligned} \qquad (12.22)$$

The approximate matrix $\mathbf{H}^{(k)}$ is adjusted on each iteration.

To understand BFGS formula, we can simply make the following approximation to the Hessian,

$$\mathbf{H}^{(k+1)} \approx \frac{\boldsymbol{\gamma}^{(k+1)} - \boldsymbol{\gamma}^{(k)}}{\mathbf{m}^{(k+1)} - \mathbf{m}^{(k)}} = \frac{\mathbf{y}_k}{\mathbf{z}_k}, \qquad (12.23)$$

which is expressed in terms of the derivative of the gradient. This is known as the secant equation:

$$\mathbf{H}^{(k+1)}\mathbf{z}_k = \mathbf{y}_k. \qquad (12.24)$$

It is straightforward to verify that the BFGS updating formula of Equation 12.21 satisfies this secant equation.

In the BFGS updating formula, the initial approximation $\mathbf{H}^{(0)}$ is often assigned to be an identity matrix. Whenever the initial approximation $\mathbf{H}^{(0)}$ is positive definite and $\mathbf{z}_k^H \mathbf{y}_k > 0$, $\mathbf{H}^{(k)}$ remains positive definite, so as to reduce the rounding errors. The approximate matrix $\mathbf{H}^{(k)}$ converges to the Hessian matrix $\mathbf{H}$ (Broyden, 1967, 1972).

12.3 Strategy for reflection seismic tomography

Figure 12.1a shows the Marmousi velocity model of size 4550×2800 m^2. We set up such a model with relatively narrow spatial extent for the purpose of testing the capacity of seismic waveform tomography using reflection data with limited source-receiver offsets, as in exploration seismology. In the frequency-domain forward modelling and tomographic inversion, the velocity model is discretised into 12.5×12.5 m^2 cells. The acoustic wave equation (12.5) is solved using a finite-difference scheme.

Figure 12.1b shows the velocity image obtained from waveform tomography, where the initial model used in the iterative inversion is a smooth version of the true velocity model, obtained by low-pass filtering. The frequencies used in inversion are within the range from 2.7 to 30 Hz with 0.3 Hz sample interval. The maximum frequency $f_{max} = 30$ Hz is due to the limitation of our frequency-domain finite-difference modelling:

$$f_{max} = \frac{v_{min}}{4h}, \qquad (12.25)$$

where the minimum velocity $v_{min} = 1500$ m/s, and cell size $h = 12.5$ m. Because the seismic reflection data are recorded with limited offsets

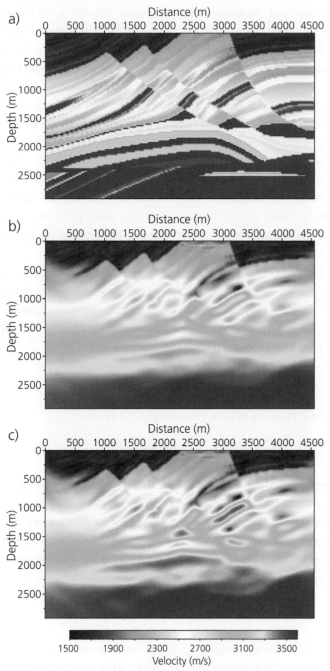

Figure 12.1 (*a*) The Marmousi velocity model. (*b*) Waveform inversion image, in which only the top portion of the model is well-reconstructed. (*c*) Final waveform tomography image, with scaled model updates during the iterative inversion.

between sources and receivers, we can only obtain a well-constructed image of the topmost portion of the model. To overcome this problem in reflection seismic waveform tomography, we propose a scheme in which depth-dependent weighting factors are applied to velocity updating in the inversion.

By definition, tomography is an integral along the ray path; that is, data residuals are spread evenly along the ray path. The residual projected to a cell is linearly related to the model update of the cell, and thus the model update depends upon the total length of ray segments within the cell, that is, the ray density. Since the ray density in the shallow depth is much higher than that in a deeper part of the model, the model update in the shallow depth is much higher than that in the deep portion of the model. This is an underlying rationale for using depth-dependent weights in model updating.

In the gradient inversion method, the objective function is minimised by updating the model vector along the opposite direction to the gradient, $\mathbf{m}^{(k+1)} = \mathbf{m}^{(k)} - \alpha \hat{\boldsymbol{\gamma}}$. Application of an inverse Hessian matrix would sharpen or focus the gradient vector $\hat{\boldsymbol{\gamma}}$ and could improve the convergence rate in an iterative solution:

$$\mathbf{m}^{(k+1)} = \mathbf{m}^{(k)} - \alpha \mathbf{H}^{-1} \hat{\boldsymbol{\gamma}}, \tag{12.26}$$

where $\mathbf{H}$ is the Hessian matrix (Equation 12.18),

$$\mathbf{H} \approx \mathrm{Re}\{\mathbf{F}^{\mathrm{H}}\mathbf{F}\}, \tag{12.27}$$

given in terms of matrix $\mathbf{F}$, the Fréchet derivatives of seismic wavefield with respect to the model elements, and its complex conjugate transpose $\mathbf{F}^{\mathrm{H}}$. Here, the data covariance matrix $\mathbf{C}_{\mathrm{d}}$ is set as an identity matrix. The diagonal of the Hessian matrix $\mathbf{H}$ is dominant compared to other components, and that its inverse matrix could be used to scale the images.

By definition, the wavefield is given by

$$u(\mathbf{r}, \omega) = s(\mathbf{r}_s, \omega) g(\mathbf{r}, \mathbf{r}_s, \omega), \tag{12.28}$$

where $\mathbf{r}$ is the position vector, $\mathbf{r}_s$ is the source position, $g(\mathbf{r}_s, \omega)$ is the source amplitude at the frequency ω, and $g(\mathbf{r}, \mathbf{r}_s, \omega)$ is Green's function in the frequency domain. In a homogeneous medium, the frequency-domain, free space, 2-D Green's function is

$$g_{2D}(\mathbf{r}, \mathbf{r}_s, \omega) = \left(\frac{-i v}{8\pi\omega R}\right)^{1/2} \exp\left(-i\omega\frac{R}{v}\right), \tag{12.29}$$

where $R = |\mathbf{r} - \mathbf{r}_s|$ is the distance from the source, and v is the acoustic velocity of the medium. Considering Equations 12.27, 12.28 and 12.29, we can find that the main diagonal elements of the inverse Hessian matrix $\mathbf{H}^{-1}$ have the following behaviour:

$$[\mathbf{H}^{-1}]_{jj} \propto R v, \tag{12.30}$$

since $F_{ij} \propto (Rv)^{-1/2}$. The analysis above leads us to a scaling scheme for the model updates in reflection waveform inversion as in the following:

$$\mathbf{x}^{(k+1)} = \mathbf{x}^{(k)} - \alpha\beta(z)\boldsymbol{\gamma}, \tag{12.31}$$

where $\beta(z)$ is a depth dependent scaling factor, $\beta(z) \propto z v(z)$. This scaling factor effectively compensates the geometrical spreading effect of the waveform amplitudes in the inversion.

Figure 12.1c is the final image of the inversion example using this scaled updating scheme. The scaling factor used is

$$\beta(z) = 1.0 + 0.2 \times 10^{-6} z v(z). \tag{12.32}$$

For depth $z = 0$, 1000, 2000, 2500 and 2750 m with corresponding average velocity $v(z) = 1500$, 2000, 2500, 3000 and 3600 m/s, we use the scale factor

$$\beta(z) = 1.0, 1.4, 2.0, 2.5 \text{ and } 3.0,$$

respectively.

If using the synthetic data with random noise, the inversion result is very close to the result for noise-free synthetic data. This is because the noise is randomly distributed in the time domain. In the frequency domain, the noise becomes ambient with a white spectrum. In other words, the frequency-domain inversion is not affected by random noise that is spectrally white. In practice, however, field seismic data contain noise that is seldom distributed randomly and hardly has a white spectrum. A group of frequencies is then used simultaneously in each iterative inversion, to suppress the noise effect in the frequency-domain inversion (Wang and Rao, 2006).

12.4 Multiple attenuation and partial compensation

We now discuss the strategies for application of waveform tomography to marine seismic data. Figure 12.2a is a sample shot record, consisting of 120 traces with a minimum source-receiver offset of 337.5 m and a maximum source-receiver offset of 1825 m. We investigate the feasibility of reflection seismic tomography within such a narrow source-receiver offset range.

The waveform tomography commonly does not include free-surface multiples, as an absorbing boundary condition at the free surface may be used in the forward calculation. Including free-surface multiples in the tomography inversion will increase the nonlinearity of the problem. As the number of multiples increases, the errors in model (and, in turn, in synthetics) will also increase. We use a narrow-offset shot record in multiple attenuation, also to avoid the wide-angle refraction of the water bottom and its multiples, as the current methodology for free-surface multiple prediction cannot properly model the refraction multiples. As marked in Figure 12.2a, the most difficult part of multiple attenuation is where the refraction wave just starts appearing. We use a data-driven multiple attenuation method, called MPI, multiple prediction through inversion (Wang, 2004b, 2007). Figure 12.2b displays the shot record after free surface multiple attenuation. The real shot record is generated by a point source, but Figure 12.2c is an equivalent line-source shot gather, after partial compensation.

Before input to waveform tomography, a shot record of real seismic data needs to be partially compensated, to become a gather that is generated from a line source. Equation 12.29 is the 2-D Green's function. For a three-dimensional (3-D) case, Green's function is

$$g_{3D}(\mathbf{r}, \mathbf{r}_s, \omega) = \frac{1}{4\pi R} \exp\left(-i\omega \frac{R}{v}\right). \tag{12.33}$$

Comparison between the 2-D and 3-D Green's functions leads to the partial compensation operator as

$$W(\omega) = \sqrt{\frac{2\pi R v}{i\omega}}. \tag{12.34}$$

In the time-domain, the operator $w(t)$ shows the following behaviour (in the far field),

$$w(t) = D_{-1/2}(t)\sqrt{2\pi R v}\,, \tag{12.35}$$

where $D_{-1/2}(t)$ is a half-integrator, defined as the inverse Fourier transform of $(i\omega)^{-1/2}$ (Deregowski and Brown, 1983). For a narrow-offset, reflection geometry, assuming $2R \propto vt$, one may obtain

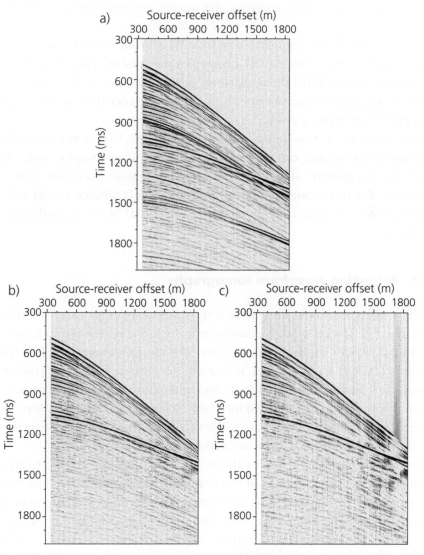

Figure 12.2 (*a*) A sample shot record with 120 traces. (*b*) The shot record after multiple attenuation. (*c*) The same shot record after partial compensation.

$$w(t) \propto D_{-1/2}(t)\, v(t)\sqrt{t} \; . \tag{12.36}$$

This partial compensation may be implemented in two steps: applying a scale factor $v(t)\sqrt{t}$ in the time domain, and multiplying the operator $(i\omega)^{-1/2}$ in the frequency domain.

Figure 12.3 displays a brute stack of the raw marine seismic data, the stack section of the seismic data set after multiple attenuation, and the same stack section after applying the partial compensation to the shot records.

Figure 12.4 closely compares a seismic trace from a point source (solid curve) and a trace after partial compensation (dashed curve), and their amplitude spectra. With the application of the operator $(i\omega)^{-1/2}$, wavelets in the trace from an equivalent line source are broader than those actually generated from a point source.

Alternatively, a two-and-a-half-dimensional (2.5-D) wave modelling and inversion scheme can be used for waveform tomography. But, as it involves integration along the infinite line-source direction that is perpendicular to the source-receiver profile, a 2.5-D scheme would take a much longer running time, compared to the 2-D wave modelling and inversion.

12.5 Reflection waveform tomography

In the reflection-seismic data, there is no significant energy recorded at low frequencies less than 6 Hz. A traveltime inversion is performed to generate the initial velocity model for the iterative waveform tomography. Referring to the stack section (Figure 12.3), traveltimes of two reflections from the pre-stack seismic data are picked, and traveltime inversion generates a layered velocity model with two interfaces (Figure 12.5). The first layer is a water layer with velocity1500 m/s. The second layer has a linear velocity of 2200 m/s at one end and 2000 m/s at distance 23 km.

In frequency-domain waveform tomography, a small group of frequencies (usually 3-5) is used simultaneously in each iterative inversion. Using a group of neighbouring frequencies in the input can suppress the noise effect in the real data, and also more data samples used in an inversion means a much better-determined inverse problem. In this example, there are 26 groups of frequencies in total in the range of 6.9–30 Hz with an interval of 0.3 Hz. The first group includes frequencies of 6.9,

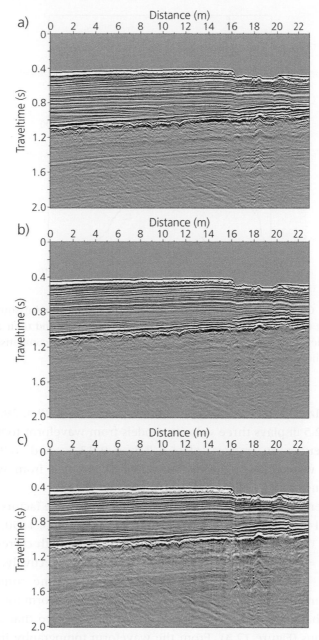

Figure 12.3 (*a*) A marine seismic section. (*b*) The section after pre-stack free-surface multiple attenuation. (*c*) The same section with amplitude and phase compensations applied to the shot records before stacking. The compensations to the shot records make the original point sources become the equivalent line sources, before they are used in waveform tomography.

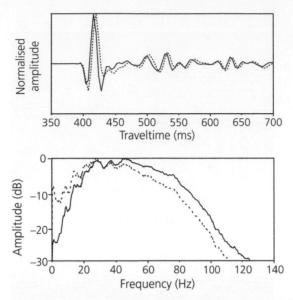

Figure 12.4 Comparison between a seismic trace from a point-source (solid curve) and the trace after partial compensation (dashed curve) and their amplitude spectra. Wavelets in a trace from a line source (i.e. after partial compensation) are broader than those from a point-source.

7.2 and 7.5 Hz, and the last group includes frequencies of 29.4, 29.7 and 30 Hz. Figure 12.5 displays three velocity models from waveform tomography using frequencies in ranges of 6.9–7.5, 6.9–13.8, and 6.9–30 Hz, respectively, where the final velocity model is obtained from waveform tomography using all frequencies in the range (6.9–30 Hz).

In the inversion example shown in Figure 12.5, the scale factors used are $\beta(z) = 1.0$, 1.14, 1.4, 1.7, and 2.0, for depth $z = 0$, 350, 800, 1150, 1500 m, respectively. These are estimated using Equation 12.23 with corresponding velocity $v(z) = 1500$, 2000, 2500, 3000 and 3300 m/s, respectively.

Although the initial model generated by traveltime tomography generates a smooth boundary for the water bottom, waveform tomography produces a sharp geometry with a spatial variation close to that shown in seismic sections (Figure 12.3). From the waveform tomography image, we can see clearly a stratified structure under the water bottom. In the second layer, at the distance between 0 and 10 km, there is a high-low-high vertical velocity variation immediately underneath the water bottom, between depths of 350 and 700 m at the left end of the profile, and a

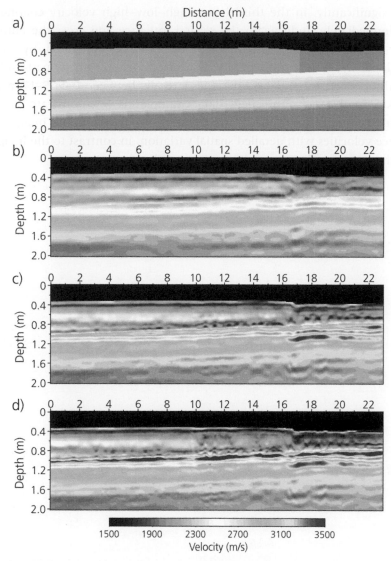

Figure 12.5 Seismic waveform tomography. (*a*) The initial velocity model, built from traveltime tomography. (*b–d*) Three velocity models reconstructed by waveform tomography, using frequencies in the ranges of 6.9–7.5, 6.9–13.8 and 6.9–30 Hz, respectively.

high–low–high–low vertical variation above the second interface. At the distance between 17 and 23 km, there is also a low–high–low vertical velocity variation within the second layer.

Most significantly, in the third layer, high–low–high velocity channels appear immediately underneath the second interface and cross the entire section. These thin-layers separation could not be generated by any conventional travel time tomography. Beneath that, the velocity pattern in the depth between 1.2 and 1.6 km varies laterally between 0–5, 5–17, and 17–23 km in distance. This field data example has demonstrated that waveform tomography is able to generate a high-resolution image of subsurface velocities with detailed spatial variation, in contrast to the image derived from conventional traveltime tomography.

CHAPTER 13

Waveform tomography with irregular topography

Seismic waveform tomography can produce a high-resolution image of the subsurface velocity model with intense spatial variation. However, as shown in the previous chapter, a flat surface is often assumed. In practice, especially for land seismic data, an irregular surface such as a mountainous area cannot be simplified to be a flat line at all. Even for a marine seismic acquisition, a rough water bottom sometimes cannot be treated as a planar interface numerically. This chapter will introduce a waveform simulation and inversion method that works for an Earth model with an irregular topography.

This chapter consists of five sections: A body-fitted grid scheme that can accurately represent an Earth model with an irregular topography; Boundary grid modification for the orthogonality, which is also necessary for properly setting any absorbing boundary condition; Pseudo-orthogonality and smoothness for an accurate waveform simulation; The acoustic wave equation and the absorbing boundary condition, which are reformulated from the physical space to the computational space; Waveform simulation and eventually tomographic inversion, to demonstrate the effectiveness of developed technology that works for irregular topographic models.

13.1 Body-fitted grids for finite-difference modelling

The procedure of tomographic inversion is to iteratively minimise the difference between the observed and the modelled wavefields, and hence

Seismic Inversion: Theory and Applications, First Edition. Yanghua Wang.
© 2017 John Wiley & Sons, Ltd. Published 2017 by John Wiley & Sons, Ltd.

an efficient and effective waveform simulation scheme is critical for iteration. Among existing technologies, finite difference methods are often used for their efficiency and simple implementation (Virieux *et al.*, 2011). In the conventional finite difference method, the Earth model is partitioned by quadrate cells with four sides perpendicular and parallel to the horizontal and vertical axes, respectively, in the Cartesian coordinate system. When there is an irregular topography, quadrate gridding will form a staircase boundary numerically, which will cause strong artificial scattering (Bleibinhaus and Rondenay, 2009). Dense grids might have some degree of improvement (Lombard *et al.*, 2008), but such an expensive approach cannot suppress numerical artefacts down to a satisfactory level.

Finite element method is also a suitable method for seismic modelling, as triangular grids can well describe an irregular topography (Zhang and Liu, 1999; Zhang, 2004). This method is computationally expensive, in comparison to the finite difference method. Käser and Igel (2001) tried triangular grids in combination with finite differencing for simulation. The errors in spatial derivative computations on unstructured grids, or irregular grids, were counteracted by using grids of higher node density. Thus, it also increased computational expense.

A curvilinear coordinate can be used to deal with an irregular topography. In curved grids, horizontal lines coincide with the interfaces and columns are still parallel to the vertical direction. An initial Cartesian model is transformed into a new computing model with a flat topography. The simulation is implemented in the curvilinear coordinate, using optimised operator of spatial derivatives (Hestholm and Ruud, 1998, 2000; Tarrass *et al.*, 2011). Whereas this method well describes an irregular topography, it needs to compute more derivatives and, thus, is more expense than a Cartesian method.

Body-fitted grid is a structured grid method, often used in hydrokinetics numerical simulation (Komatitsch *et al.*, 1996). It is also a curved grid method producing pseudo-orthogonal meshes. Grids are quadrangle and, as structured, keep the similar neighbourhood relationships as they in Cartesian coordinates. By mapping curved grids onto rectangular grids in computational space, the basic finite difference scheme can be adopted straightforwardly.

Simple body-fitted grids can be generated by interpolation, which is an efficient method but strongly relies on the initial control points in the boundary. The surface curve is represented by a polynomial, and any singular cells generated will crush computation. Zhang and Chen (2006) used a finite element software to generate body-fitted grids, for finite difference modelling. On the other hand, Thomas and Middlecoeff (1980), Thompson *et al.* (1985) and Hoffman and Chiang (1993) presented a sophisticated method that solves a hyperbolic system of equations with appropriate control functions. Because of improved orthogonality and smoothness with this grid method, numerical diffusion can be reduced and better accuracy is achieved in waveform simulation.

The initial meshes can be set up reliably by linear interpolation, and then body-fitted grids may be generated by solving an elliptic system of equations. When a study area is encircled by artificial boundaries, grid points in four-side boundaries and connection zones should be modified. This modification will dramatically improve the orthogonality. Based on these structured or regular grids, the acoustic wave equation and the absorbing boundary condition will be reformulated in the computational space. Waveform simulation is implemented in the frequency domain using a finite difference scheme.

Given a model in the physical space with coordinates (x, z), computation is implemented in a space with coordinates (ξ, η). The relation between the physical space and the computational space is given by the following group of equations:

$$\frac{\partial^2 \xi}{\partial x^2} + \frac{\partial^2 \xi}{\partial z^2} = M(\xi, \eta) \ ,$$

$$\frac{\partial^2 \eta}{\partial x^2} + \frac{\partial^2 \eta}{\partial z^2} = N(\xi, \eta) ,$$

(13.1)

where $M(\xi, \eta)$ and $N(\xi, \eta)$ are two terms controlling the rate of grid spacing changes in both directions. These partial differential equations of elliptic type are called Poisson's equation. To find the physical space coordinates $x = x(\xi, \eta)$ and $z = z(\xi, \eta)$ corresponding to any rectangular cells in the computational space, Equation 13.1 may be transferred to

$$\alpha \frac{\partial^2 x}{\partial \xi^2} - 2\beta \frac{\partial^2 x}{\partial \xi \partial \eta} + \gamma \frac{\partial^2 x}{\partial \eta^2} + J^2 \left(M \frac{\partial x}{\partial \xi} + N \frac{\partial x}{\partial \eta} \right) = 0 \ ,$$

$$\alpha \frac{\partial^2 z}{\partial \xi^2} - 2\beta \frac{\partial^2 z}{\partial \xi \partial \eta} + \gamma \frac{\partial^2 z}{\partial \eta^2} + J^2 \left(M \frac{\partial z}{\partial \xi} + N \frac{\partial z}{\partial \eta} \right) = 0 \ ,$$

(13.2)

where

$$\alpha = \dot{x}_\eta^2 + \dot{z}_\eta^2 , \qquad \beta = \dot{x}_\xi \dot{x}_\eta + \dot{z}_\xi \dot{z}_\eta ,$$

$$\gamma = \dot{x}_\xi^2 + \dot{z}_\xi^2 , \qquad J \equiv \frac{\partial(x,z)}{\partial(\xi,\eta)} = \dot{x}_\xi \dot{z}_\eta - \dot{z}_\xi \dot{x}_\eta . \tag{13.3}$$

To solve the discretised Poisson equation iteratively, quantities $\dot{x}_\xi = \partial x / \partial \xi$, $\dot{z}_\xi = \partial z / \partial \xi$, $\dot{x}_\eta = \partial x / \partial \eta$ and $\dot{z}_\eta = \partial z / \partial \eta$ are assumed to be known, evaluated based on the current solution.

In seismic waveform simulation, we set $M(\xi,\eta) = 0$, and evaluate $N(\xi,\eta)$ in the depth direction, so that the grid can be sparse in the deep area with high velocity and can be tight-knit in the shallow area with low velocity. The controlling term is evaluated by (Rao and Wang, 2013)

$$N(\xi,\eta) = -\frac{2L(\alpha+\gamma)\Delta r}{J^2 \dot{z}_\eta (1+r)^2} , \tag{13.4}$$

where $L = z_{i,j+1} - z_{i,j-1}$ is the closest distance in the η direction between two points next to point $z_{i,j}$, and $r = (z_{i,j+1} - z_{i,j})/(z_{i,j} - z_{i,j-1})$ is a dimensionless grid ratio. Associating with a general trend of velocity variation in depth, a grid ratio can be changed from r to $r + \Delta r$.

Any grid points should satisfy the following orthogonality condition,

$$\beta = \dot{x}_\xi \dot{x}_\eta + \dot{z}_\xi \dot{z}_\eta = 0 . \tag{13.5}$$

Therefore, a measurement for the quality of computational grids is

$$Q = \sum_{i,j} \beta(i,j) , \tag{13.6}$$

where (i,j) are indexes of a grid. A small Q value means a better orthogonality of meshes and, ideally, $Q \to 0$.

13.2 Modification of boundary points

Solving Poisson's equation (13.2) generates interior grids that usually have excellent spatial distribution. However, grids at boundaries are not orthogonal yet, and we need to modify the position of boundary points (Patantonis and Atharassiadis, 1985), following the pragmatic rule $\beta = 0$. Meanwhile, the smoothness in both ξ and η directions should be also considered after boundary point modification. While orthogonality in boundary grids is necessary for properly setting any absorbing boundary condition for waveform simulation, smoothness is critical to the accuracy of waveform simulation. Therefore, a modified boundary point should satisfy the following system of three equations:

$$\begin{cases} (x_{i+1,j} - x_{i-1,j})(\tilde{x}_{i,j+1} - x_{i,j}) + (z_{i+1,j} - z_{i-1,j})(\tilde{z}_{i,j+1} - z_{i,j}) = 0, \\ (x_{i+1,j+1} - \tilde{x}_{i,j+1})(\tilde{z}_{i,j+1} - z_{i-1,j+1}) - (\tilde{x}_{i,j+1} - x_{i-1,j+1})(z_{i+1,j+1} - \tilde{z}_{i,j+1}) = 0, \\ (x_{i,j+2} - \tilde{x}_{i,j+1})(\tilde{z}_{i,j+1} - z_{i,j}) - (\tilde{x}_{i,j+1} - x_{i,j})(z_{i,j+2} - \tilde{z}_{i,j+1}) = 0. \end{cases}$$

$$(13.7)$$

This system modifies grid point $(i, j+1)$ from $(x_{i,j+1}, z_{i,j+1})$ to $(\tilde{x}_{i,j+1}, \tilde{z}_{i,j+1})$. The first equation is from the orthogonality equation (13.5), and the rest represent the smoothness in ξ and η directions respectively at $(x_{i,j+1}, z_{i,j+1})$. In practice, the modification can be compromised by

$$\hat{x}_{i,j+1} = \frac{w\tilde{x}_{i,j+1} + x_{i,j+1}}{1+w},$$

$$\hat{z}_{i,j+1} = \frac{w\tilde{z}_{i,j+1} + \tilde{z}_{i,j+1}}{1+w},$$

$$(13.8)$$

where w is a parameter set to keep it smooth enough after modification.

In summary, the procedure for generating body-fitted grids consists of the following steps:

1) Setting evenly spaced initial points on four boundaries.
2) Generating initial internal grids by linear interpolation over those initial points on boundaries.
3) Solving Poisson's equation for orthogonal internal meshes.
4) Modifying boundary points.
5) Measuring the quality of grids.
6) Adjusting $M(\xi, \eta)$ and $N(\xi, \eta)$, and going to step 3 to repeat the computation.

The iteration is stopped when the Q value is sufficiently small.

13.3 Pseudo-orthogonality and smoothness

Figure 13.1a is a simple example in which the bottom boundary and the left and right boundaries are planar, and the top topography is an analytical curve defined by a Gaussian function of the lateral coordinate,

$40\exp[-(0.01x-1.6)^2]$. In body-fitted grids, generated from Poisson's equation, most internal meshes are orthogonal, except for near-boundary zones (Figure 13.1a).

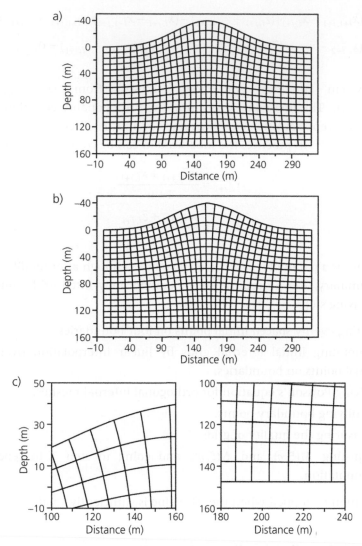

Figure 13.1 (*a*) Body-fitted grids (without boundary point modification). (*b*) Body-fitted grids with boundary-point modification. (*c*) Zoomed-in meshes between $x = [100, 160]$ m, and $z = [-10, -50]$ m, and between $x = [180, 240]$ m and $z = [100, 160]$ m. Because of boundary-point modification, both boundary and internal grids have a good orthogonal performance.

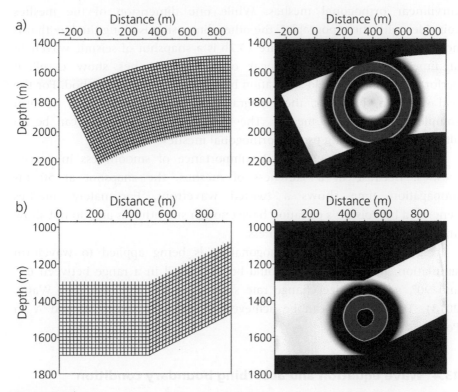

Figure 13.2 (*a*) Seismic wave simulation in a homogeneous fan area: the orthogonal meshes to partition the study area, plotted by each 5 grids, and a snapshot of the wavefield at a time 80 ms. (*b*) Wavefield simulation in a homogeneous area with a skewness: the meshes to partition the area, and a snapshot at 50 ms, showing a twisted wavefront because of the skewness of meshes.

These unorthogonal meshes will twist the wavefield in simulation. Therefore, in the iterative mesh generation procedure, we shall modify these boundary points as well. The resultant meshes, Figure 13.1b, show smooth and orthogonal characters in both internal and boundary grids. Figure 13.1c consists of zoomed-in pictures of the top boundary at range $x = [100, 160]$ m and $x = [-10, -50]$ m, and the bottom boundary at $x = [180, 240]$ m and $z = [100, 160]$ m. Iterative boundary-point modification and inner mesh refining can make meshes have a better orthogonal behaviour that is suitable for a finite-difference calculation.

Figure 13.2a is a sample of curvilinear grids with which to partition a fan area. As the topography is defined by circular-arc function, it can form ideal curvilinear orthogonal meshes. While one dimension of the meshes coincides with the topography, the other dimension is along the line that is normal to the topography. Figure 13.2b is a snapshot of seismic wavefield at time 80 ms. The curvilinear orthogonal meshes show excellent performance in wavefield simulation by a finite difference method. For the real situation, however, the topography is often too complex to form completely orthogonal meshes. The body-fitted grid method can be an alternative for building pseudo-orthogonal meshes.

Figure 13.2b demonstrates the importance of smoothness in curved meshes. Because of a skewness of meshes, the snapshot at 50 ms propagation time shows a twisted wavefront. Fortunately, meshes generated by Poisson's equation have enough smoothness and do not show any of this kind of sudden skewness.

In general, for pseudo-orthogonal grids being applied to waveform simulation, the acute angle should be controlled in a range between 67° and 90°, and grid-size change rate should be $\Delta r < 5\%$ (Rao and Wang, 2013). These targets can be achieved by adjusting terms M and N in Poisson's equation.

13.4 Wave equation and absorbing boundary condition

The relation between physical coordinates (x, z) and computational coordinates (ξ, η) may be established through the first-order differentiation,

$$\frac{\partial u}{\partial x} = \xi_x \frac{\partial u}{\partial \xi} + \eta_x \frac{\partial u}{\partial \eta} \,, \tag{13.9}$$

and the second-order differentiation,

$$\frac{\partial^2 u}{\partial x^2} = \dot{\xi}_x^2 \frac{\partial^2 u}{\partial \xi^2} + \dot{\eta}_x^2 \frac{\partial^2 u}{\partial \eta^2} + 2\dot{\xi}_x \dot{\eta}_x \frac{\partial^2 u}{\partial \xi \partial \eta} + \ddot{\xi}_{xx} \frac{\partial u}{\partial \xi} + \ddot{\eta}_{xx} \frac{\partial u}{\partial \eta} \,. \tag{13.10}$$

Then, the acoustic wave equation

$$\left(\nabla^2 + \frac{\omega^2}{v^2(x,z)} \right) u(x,z) = 0 \,, \tag{13.11}$$

becomes

$$(\dot{\xi}_x^2 + \dot{\xi}_z^2)\frac{\partial^2 u}{\partial \xi^2} + (\dot{\eta}_x^2 + \dot{\eta}_z^2)\frac{\partial^2 u}{\partial \eta^2} + 2(\dot{\xi}_x\dot{\eta}_x + \dot{\xi}_z\dot{\eta}_z)\frac{\partial^2 u}{\partial \xi \partial \eta}$$

$$+ (\ddot{\xi}_{xx} + \ddot{\xi}_{zz})\frac{\partial u}{\partial \xi} + (\ddot{\eta}_{xx} + \ddot{\eta}_{zz})\frac{\partial u}{\partial \eta} + \frac{\omega^2}{v^2}u = 0 \,. \tag{13.12}$$

This is the wave equation in the computational space.

Then, a proper absorbing boundary condition is critical in the case with an irregular topography, since a small incidence angle in a flat boundary could be a big angle in an irregular boundary. Consider a plane wave as

$$u = u_0 \exp[\mathrm{i}\,(\omega t - k_x x)]\,, \tag{13.13}$$

where t is traveltime, ω is the angular frequency, and k_x is the wavenumber in x direction. The aim of an absorbing boundary condition is to modify the wave solution in the numerical computation such that the amplitude is attenuated to

$$\tilde{u} = u \exp(-\alpha x)\,, \tag{13.14}$$

where α is an attenuation coefficient and is chosen such that $e^{-\alpha x}\,|_{x=0} = 1$ and $e^{-\alpha x}\,|_{x>0} < 1$. That is, at position $x = 0$ the solution is perfectly matched (before finite-difference approximation) and will not cause any reflection. This is the perfectly matched layer (PML) method, developed first for electromagnetic waves (Bérenger, 1994, 1996). To understand the concept, we put it in the context of acoustic wave propagation, and reformulate it for the case with an irregular topography in the computational space.

Defining the attenuation coefficient as a frequency-dependent function,

$$\alpha x = \frac{k_x}{\omega} \int_0^x d(\ell)\mathrm{d}\ell\,, \tag{13.15}$$

where $d(x)$ is a damping factor within the PML region. Then

$$\tilde{u} = u_0 \exp[\mathrm{i}(\omega t - (k_x - \mathrm{i}\alpha)x)]$$

$$= u_0 \exp\left[\mathrm{i}\left(\omega t - k_x\left(x + \frac{1}{\mathrm{i}\omega}\int_0^x d(\ell)\mathrm{d}\ell\right)\right)\right] \tag{13.16}$$

$$= u_0 \exp[\mathrm{i}(\omega t - k_x\tilde{x})]\,,$$

where

$$\tilde{x} = x + \frac{1}{i\omega} \int_0^x d(\ell)\, d\ell \, . \tag{13.17}$$

Therefore, PML involves the change of a real-valued spatial variable x to a complex-valued variable $\tilde{x}$, as defined by Equation 13.17.

Changing variable $x \to \tilde{x}$ is equivalent to the following change to partial differential:

$$\frac{\partial}{\partial x} \to \frac{1}{s_x} \frac{\partial}{\partial x}, \tag{13.18}$$

where s_x is a complex stretching factor. According to Equation 13.17, the complex stretching factor is defined by

$$s_x \equiv \frac{\partial \tilde{x}}{\partial x} = 1 + \frac{d(x)}{i\omega}\, . \tag{13.19}$$

To stretch the computational coordinates $\xi \to \tilde{\xi}$ and $\eta \to \tilde{\eta}$, partial differentials in Equation 13.12 are replaced as

$$\begin{aligned}
\frac{\partial}{\partial \xi} &\to \frac{1}{s_\xi} \frac{\partial}{\partial \xi}, \\[1ex]
\frac{\partial}{\partial \eta} &\to \frac{1}{s_\eta} \frac{\partial}{\partial \eta}, \\[1ex]
\frac{\partial^2}{\partial \xi^2} &\to \frac{\partial^2}{\partial \tilde{\xi}^2} = \frac{1}{s^3(\xi)} \frac{1}{i\omega} \frac{\partial d(\xi)}{\partial \xi} \frac{\partial}{\partial \xi} + \frac{1}{s^2(\xi)} \frac{\partial^2}{\partial \xi^2}, \\[1ex]
\frac{\partial^2}{\partial \eta^2} &\to \frac{\partial^2}{\partial \tilde{\eta}^2} = \frac{1}{s^3(\eta)} \frac{1}{i\omega} \frac{\partial d(\eta)}{\partial \eta} \frac{\partial}{\partial \eta} + \frac{1}{s^2(\eta)} \frac{\partial^2}{\partial \eta^2}\, .
\end{aligned} \tag{13.20}$$

Then, the acoustic wave equation in the PML absorbing zone is written as

$$\begin{aligned}
&\frac{(\dot{\xi}_x^2 + \dot{\xi}_z^2)}{i\omega\, s^3(\xi)} \frac{\partial d(\xi)}{\partial \xi} \frac{\partial u}{\partial \xi} + \frac{(\dot{\xi}_x^2 + \dot{\xi}_z^2)}{s^2(\xi)} \frac{\partial^2 u}{\partial \xi^2} + \frac{(\dot{\eta}_x^2 + \dot{\eta}_z^2)}{i\omega\, s^3(\eta)} \frac{\partial d(\eta)}{\partial \eta} \frac{\partial u}{\partial \eta} \\[1ex]
&+ \frac{(\dot{\eta}_x^2 + \dot{\eta}_z^2)}{s^2(\eta)} \frac{\partial^2 u}{\partial \eta^2} + 2(\dot{\xi}_x \dot{\eta}_x + \dot{\xi}_z \dot{\eta}_z) \frac{1}{s_\xi s_\eta} \frac{\partial^2 u}{\partial \xi \partial \eta} + \frac{(\ddot{\xi}_{xx} + \ddot{\xi}_{zz})}{s_\xi} \frac{\partial u}{\partial \xi} \\[1ex]
&+ \frac{(\ddot{\eta}_{xx} + \ddot{\eta}_{zz})}{s_\eta} \frac{\partial u}{\partial \eta} + \frac{\omega^2}{v^2} u = 0\, .
\end{aligned} \tag{13.21}$$

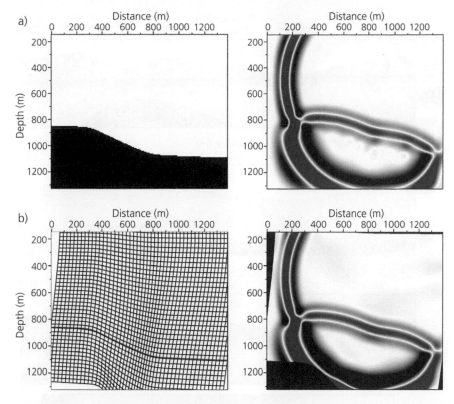

Figure 13.3 (*a*) Velocity model with staircase boundary, caused by quadrate grids partition at a dip subsurface boundary, and the corresponding snapshot (at 190 ms), showing that the staircase boundary could cause a strong scattering effect in the wavefield. (*b*) Pseudo-orthogonal grid for the same area, where the dark line is the subsurface boundary, and the corresponding snapshot (at 190 ms) without scattering effect, when using pseudo-orthogonal grids.

In the conventional finite-difference scheme using quadrate grids in the Cartesian coordinate, a 'staircase' boundary will cause strong scattering in the wavefield (Figure 13.3a). It can be improved by fine partitioning with increased computational expense.

In contrast, when using pseudo-orthogonal grids for the same area, there is no extra computational expense for describing the irregular boundary with finite differencing, and there is no artificial scattering effect (Figure 13.3b). With pseudo-orthogonal grids, there is a simple relationship between grid connections and, thus, it is also easy to solve the boundary condition.

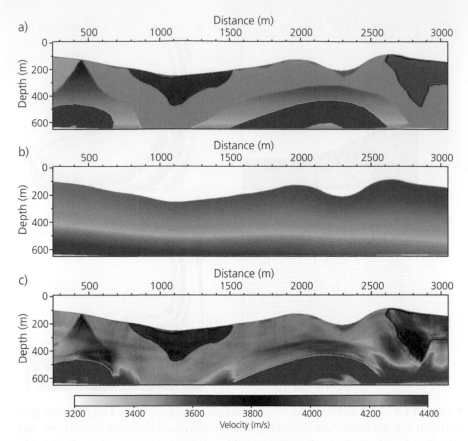

Figure 13.4 (*a*) A complicated velocity model with irregular topography. (*b*) The initial velocity model for waveform inversion. (*c*) Reconstructed velocity model generated by waveform tomography.

13.5 Waveform tomography with irregular topography

This section demonstrates waveform simulation and ultimately tomographic inversion, for a near-surface velocity model with an irregular topography.

Figure 13.4a is a complex velocity model. The top boundary is a smooth curved surface, with the highest depth difference of about 100 m. A line of sources and a line of receivers are placed along the top boundary with 10 m interval in the horizontal direction.

The body-fitted grid method is used to partition this model into 670×150 grids. The finite-difference scheme is applied to approximate the

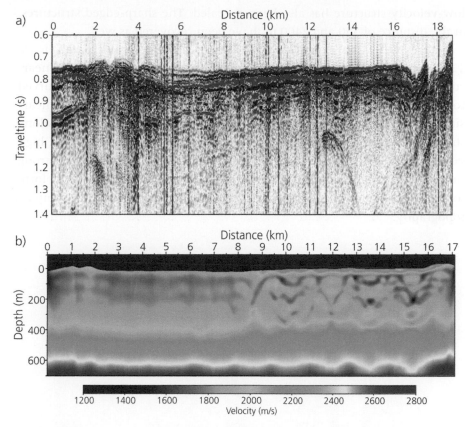

Figure 13.5 (*a*) A constant-offset profile from a complex near-surface land seismic data set. (*b*) The velocity model reconstructed by frequency-domain waveform tomography, using frequency components between 10 and 31.4 Hz.

acoustic wave equation and the PML boundary condition in computational space. A synthetic seismic data set is generated by using a synthetic source of the Ricker wavelet with a 30 Hz dominant frequency.

For waveform tomography, an initial model with low-wavenumber components (Figure 13.4b) is set as a smoothed version of the true model (Figure 13.4a). A reconstructed velocity model (Figure 13.4c) by the frequency domain waveform tomography (described in the previous chapter) clearly recovers main structures in the true model, especially the low-velocity structures with irregular triangle forms, both at the distance of 900–1700 m and the depth of 200–300 m, and the distance of 2700–3650 m and the depth of 100–500 m. The thin high-velocity layer on the top of a low-velocity structure has also been revealed. The sharp-edged structures, at the distance of 550 and 2400 m, have been recovered as well.

Figure 13.5 is an example of application to field seismic data. Figure 13.5a is a constant-offset profile of the field seismic line with complex near-surface geology. Figure 13.5b shows a velocity model with curved topography, reconstructed by waveform tomography. In the left half, near-surface velocity appears as layered structure, and generally increases along the depth. In the right half, the top-most thin skin has a modest high velocity. And immediately beneath that, the velocity model shows strong heterogeneity. This example demonstrates the effectiveness of waveform tomography for field seismic data, acquired from areas with complex topography.

CHAPTER 14

Waveform tomography for seismic impedance

Seismic waveform tomography, which is a wave equation-based inversion procedure, conventionally attempts to reconstruct the subsurface velocity distribution, as presented in the previous two chapters. However, seismic reflection is a direct response to the impedance contrast, not to the velocity variation. Reflection seismic waveform tomography would be an inversion technology used to invert for ideally the impedance parameter, not the velocity parameter.

Commonly used impedance inversion methods are based on the convolutional model, which assumes a seismic trace being an Earth reflectivity series convolved with a wavelet. Therefore, these impedance inversion methods generally involve two steps: the reflectivity inversion and then the impedance inversion (Chapters 8 and 9). There are rarely any studies on a wave equation-based impedance inversion. The latter would be much more complicated and difficult mathematically than a convolutional model-based inversion. However, as the wave equation is capable to precisely describe seismic wave propagation through the subsurface media, a wave-equation based inversion procedure would be able to reconstruct the subsurface impedance image with a much higher resolution than the convolutional model-based inversion.

A possible wave equation-based inversion procedure can be to generate the impedance image indirectly, by first inverting for impedance-related elastic parameters and then transforming them into the impedance. For example, an inversion simultaneously inverts for the velocity and density, or for the Lamé parameters (λ, μ) and density. These elastic parameters

Seismic Inversion: Theory and Applications, First Edition. Yanghua Wang.
© 2017 John Wiley & Sons, Ltd. Published 2017 by John Wiley & Sons, Ltd.

will form the impedance. One can also reformulate the wave equation to a form containing the impedance and density parameters, and then use this newly formed wave equation in an inversion to invert for the impedance and density, simultaneously or sequentially (Tarantola, 1986).

These inversion methods mentioned above involve a simultaneous inversion of at least two parameters. Considering sensitivities of the objective function, with respect to different physical parameters, and also the influence of a potential coupling effect among those parameters, it might be difficult to obtain a satisfactory multi-parameter inversion result. It in turn cannot guarantee a sufficient accuracy of the impedance parameter derived from the result.

One of the possible ways to directly invert for the impedance parameter is to set the velocity variation to be known and to invert for the impedance parameter. If the background velocity contains the correct low wavenumbers and can accurately model the kinematics of seismic reflections, reflection seismic waveform inversion can be used to reconstruct the acoustic impedance image (Plessix and Li, 2013).

In this chapter, we reformulate the ordinary wave equation, which contains the density and velocity parameters, to an equation defined in terms of only the impedance parameter, and then apply the waveform tomography mode for the impedance inversion. Santosa and Schwetlick (1982) once formulated the shear wave equation with shear impedance and inverted for the shear impedance using a characteristic method. In this chapter, we develop a full waveform inversion method to invert for the acoustic impedance from post-stack zero-offset seismic traces.

For the inversion of zero-offset seismic data, a one-dimensional (1-D) wave equation can be used. This 1-D wave equation is reformulated to contain only the impedance parameter. In order to maintain lateral continuity in the impedance image, a natural choice would be a multichannel inversion scheme that utilises all of seismic traces simultaneously, instead of a simple trace by trace implementation. This chapter proposes a strategy that can use only a few of traces, collected sparsely from the space, to reconstruct the entire two-dimensional (2-D) impedance model at the same time. The core technique of this strategy is to parameterise the subsurface impedance model by a truncated Fourier series. When the inversion seeks for the Fourier coefficients, any coefficient update will affect the entire impedance model.

This parameterisation scheme with a Fourier series was applied first to seismic traveltime and amplitude inversion for the interface geometry and the velocity variation, as presented in Chapter 11. One of the advantages is an easy implementation of a multi-scale inversion: inverting first for the coefficients of low-wavenumber components, which represent the background of the impedance model, and then for the coefficients of high-wavenumber components, which represent details of the impedance variation. Because the inversion utilises all seismic traces and guarantees the spatial continuity in the resultant image, this impedance inversion method has proved to be efficient and effective.

14.1 Wave equation and model parameterisation

The 1-D acoustic wave equation is

$$\frac{\partial}{\partial z}\left(\kappa(z)\frac{\partial u(z,t)}{\partial z}\right) - \rho(z)\frac{\partial^2 u(z,t)}{\partial t^2} = 0 , \qquad (14.1)$$

where z is the depth, $\kappa(z) = \rho(z)v^2(z)$ is the bulk modulus, $\rho(z)$ is the density, $v(z)$ is the velocity, $u(z,t)$ is the wavefield of displacement, and t is the wave travelling time. Set a new 'depth' variable:

$$\tau = 2\int_0^z \frac{ds}{v(s)} , \qquad (14.2)$$

where τ is the vertical two-way traveltime. It may be referred to as depth-time, with the units of time. This coordinate setting will transform the physical model from the true depth coordinate to the vertical time domain. Then, the original acoustic wave equation becomes

$$4\frac{\partial}{\partial \tau}\left(Z(\tau)\frac{\partial u(\tau,t)}{\partial \tau}\right) - Z(\tau)\frac{\partial^2 u(\tau,t)}{\partial t^2} = 0 , \qquad (14.3)$$

where $Z(\tau)$ is the acoustic impedance along depth-time τ. This equation contains only a single elastic parameter, the impedance parameter $Z = \rho v = \kappa / v$, and can be solved numerically using a finite difference method.

Figure 14.1 displays the 2-D Marmousi impedance model, presented in both the depth domain and the depth-time domain. When the depth is transferred to the vertical time τ, smoothly variable interfaces could be severely skewed. This depth-time domain Marmousi model is discretised into 180×174 cells (in the x and τ direction, respectively).

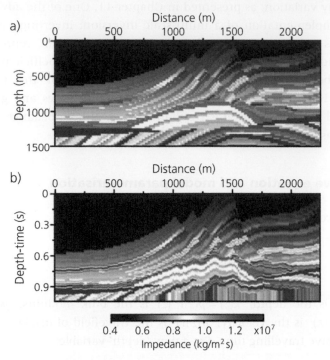

Figure 14.1 (*a*) The Marmousi impedance model in the depth domain. (*b*) The same Marmousi model, but transformed to the depth-time domain.

For the inversion, the 2-D impedance model is parameterised by the following truncated Fourier series:

$$Z(x,\tau) = \sum_{(\ell,n)=(0,0)}^{(L,N)} [a_{\ell n} \cos(\ell k_0 x) \cos(n\omega_0 \tau) + b_{\ell n} \cos(\ell k_0 x) \sin(n\omega_0 \tau)$$

$$+ c_{\ell n} \sin(\ell k_0 x) \cos(n\omega_0 \tau) + d_{\ell n} \sin(\ell \Delta k_0 x) \sin(n\omega_0 \tau)],$$

$$(14.4)$$

where $Z(x,\tau)$ is the impedance parameter, $\{a,b,c,d\}$ with subscripts ℓn are the amplitude coefficients of the harmonic terms, (L,N) are the number of harmonic terms in each dimension, k_0 is the fundamental wavenumber in the horizontal space domain, and ω_0 is the fundamental frequency in the depth-time domain. In this chapter, we set $k_0 = \pi / X$ and $\omega_0 = \pi / T$, where X and T are the durations in the x and τ direction, respectively.

Using this model parameterisation, the impedance inversion becomes an inversion for the Fourier coefficients. The 2-D impedance model can be reconstructed directly based on inverted Fourier coefficients.

14.2 The impedance inversion method

In waveform tomography, the objective function is defined in terms of data misfit, as

$$\phi(\mathbf{m}) = \frac{1}{2} \sum_p \| \mathbf{u}_{obs}^{(p)} - \mathbf{u}^{(p)}(\mathbf{m}) \|^2 , \qquad (14.5)$$

where $\mathbf{u}_{obs}^{(p)} \equiv \mathbf{u}_{obs}(p\Delta x)$ represents a single seismic trace at the lateral position $p\Delta x$, $\mathbf{m}$ is the model vector consisting of the Fourier coefficients, and $\mathbf{u}^{(p)}(\mathbf{m})$ is the synthetic counterpart, calculated based on the current model vector $\mathbf{m}$. Waveform tomography in this chapter is implemented in the time domain.

For calculating the gradient vector $\boldsymbol{\gamma}$, let us set an objective function for each individual trace as

$$\phi^{(p)}(\mathbf{m}) = \frac{1}{2} \| \mathbf{u}_{obs}^{(p)} - \mathbf{u}^{(p)}(\mathbf{m}) \|^2 . \qquad (14.6)$$

Then the objective function in Equation 14.5 becomes $\phi(\mathbf{m}) = \sum_p \phi^{(p)}(\mathbf{m})$, as we deal with the 1-D model in a multichannel fashion.

The first-order derivative of this individual objective function $\phi^{(p)}(\mathbf{m})$, with respect to the coefficient $a_{\ell n}$, is

$$\frac{\partial \phi^{(p)}(\mathbf{m})}{\partial a_{\ell n}} = - \left[\frac{\partial \mathbf{u}^{(p)}(\mathbf{m})}{\partial a_{\ell n}} \right]^{\mathrm{T}} \Delta \mathbf{u}^{(p)} , \qquad (14.7)$$

where $\Delta \mathbf{u}^{(p)}$ is the residual vector, and $\partial \mathbf{u}^{(p)} / \partial a_{\ell n}$ is a column vector of the Fréchet matrix, which can be calculated through

$$\frac{\partial \mathbf{u}^{(p)}(\mathbf{m})}{\partial a_{\ell n}} = \sum_h \left(\frac{\partial Z_h}{\partial a_{\ell n}} \right)^{(p)} \frac{\partial \mathbf{u}^{(p)}}{\partial Z_h} , \qquad (14.8)$$

where $Z_h \equiv (h\Delta\tau, p\Delta x)$ is the 1-D impedance variable associated with the specific pth trace, and $\partial Z_h / \partial a_{\ell n} = \cos(\ell k_0 p\Delta x)\cos(n\omega_0 h\Delta\tau)$. Then, Equation 14.7 becomes

$$\frac{\partial \phi^{(p)}(\mathbf{m})}{\partial a_{\ell n}} = \sum_h \left(\frac{\partial Z_h}{\partial a_{\ell n}} \right)^{(p)} \frac{\partial \phi^{(p)}}{\partial Z_h}, \tag{14.9}$$

where

$$\frac{\partial \phi^{(p)}}{\partial Z_h} = -\left[\frac{\partial \mathbf{u}^{(p)}}{\partial Z_h} \right]^{\mathrm{T}} \Delta \mathbf{u}^{(p)}. \tag{14.10}$$

Equation 14.10 is the first derivative of the individual objective function, with respect to the impedance Z_h. It can be calculated by cross-correlation between two wavefields:

$$\frac{\partial \phi^{(p)}}{\partial Z(\tau)} = \int_0^T \left(4 \frac{\partial u_0(\tau,t)}{\partial \tau} \frac{\partial u_{\mathrm{b}}(\tau,t)}{\partial \tau} - \frac{\partial u_0(\tau,t)}{\partial t} \frac{\partial u_{\mathrm{b}}(\tau,t)}{\partial t} \right) dt, \tag{14.11}$$

where $u_0(\tau,t)$ is the forward wavefield generated with the normal seismic sources, and $u_{\mathrm{b}}(\tau,t)$ is the back-propagation wavefield, generated with the data residuals. The derivation is summarised in Appendix D. For detailed references, please refer to Tarantola (1984) and Gauthier *et al.* (986).

Once $[\partial \phi^{(p)} / \partial Z]_{h\Delta\tau}$ is obtained, Equation 14.9 provides the first derivative of the individual objective function, with respect to the Fourier coefficients. Then, an element of the gradient vector $\boldsymbol{\gamma}$ of the full objective function, with respect to the Fourier coefficient $a_{\ell n}$, is

$$\frac{\partial \phi(\mathbf{m})}{\partial a_{\ell n}} = \sum_p \frac{\partial \phi^{(p)}(\mathbf{m})}{\partial a_{\ell n}}. \tag{14.12}$$

The derivative with respect to other coefficients are calculated in the same way.

The Fourier coefficients can be updated by

$$\mathbf{m}^{(k+1)} = \mathbf{m}^{(k)} - \alpha \boldsymbol{\gamma}, \tag{14.13}$$

where α is the step length. Section 12.2 has introduced an optimal method using the inverse Hessian matrix, approximated by the BFGS technology. However, when the Hessian needs a huge storage, the step length is selected using the linear search method, which tries three α values, fits the corresponding data misfits with a hyperbolic function, and finds an α value associated with the minima (Nocedal and Wright, 2006; Vigh *et al.*, 2012).

14.3 Inversion strategies and the inversion flow

Exploiting the advanced features of the model parameterisation, the following inversion strategies may be adopted in the inversion.

Strategy 1: Multi-scale inversion reconstructed from low to high wavenumber components

Smaller values of (L, N) in Equation 14.4 correspond to the low wavenumber components, which represent the background information of the model. Larger values of (L, N) correspond to high wavenumber components, representing the details of the model. Figure 14.2 displays a series of 2-D impedance models reconstructed with different numbers of harmonic terms. When the numbers of (L, N) are gradually increased, more details of the model variations are recovered.

The difference between the true model $\mathbf{Z}_{\text{true}}$ and the reconstructed model $\mathbf{Z}$ is evaluated by $\| \mathbf{Z} - \mathbf{Z}_{\text{true}} \| / \| \mathbf{Z}_{\text{true}} \|$, where $\| \cdot \|$ is the L_2 norm of a vector. The errors between the true model and various reconstructed models are listed as the following:

$L \times N$	10×10	20×20	40×40	80×80
Errors	0.120	0.098	0.067	0.039

It is obvious that, the larger the L and N are, the smaller the error will be.

According to the sensitivity analysis, traveltimes are more sensitive to the low wavenumber components of the model (Wang and Pratt, 1997). Therefore, during the inversion, when starting from smaller (L, N) in the model, the inversion generates synthetic seismic traces that have an accurate kinematic feature. When moving on to larger (L, N) in the model, the inversion matches synthetics to the field data for more accurate amplitudes. It reveals the model details gradually.

Strategy 2: Seismic trace grouping

When inverting a 2-D seismic profile, we do not invert all of the data at the same time. Instead, we resample the data spatially into smaller groups and each group nearly covers the entire computation region. As shown in Figure 14.3, a layout of trace positions of a 2-D seismic profile, we first invert the seismic traces labelled with blue dots, then move on to invert the

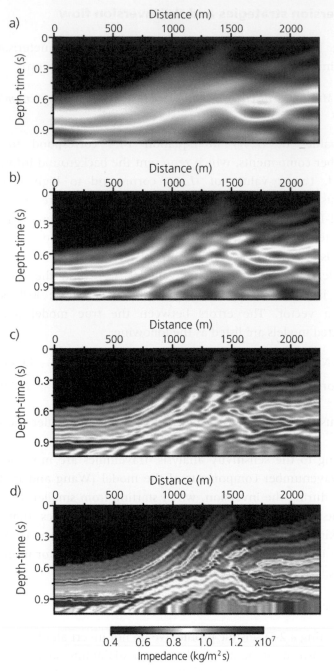

Figure 14.2 The 2-D impedance models, represented by truncated Fourier series with different numbers of harmonic terms: (*a*) $(L, N) = (10, 10)$, (*b*) $(L, N) = (20, 20)$, (*c*) $(L, N) = (40, 40)$, and (*d*) $(L, N) = (80, 80)$.

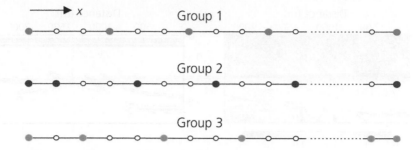

Figure 14.3 Grouping traces of a 2-D seismic profile, as an input to the impedance inversion. The first group is labelled with blue dots, the second group is labelled with red dots, and the third group is labelled with green dots. For each group, traces of the two ending points must be included, to satisfy the periodic condition of the Fourier series.

seismic traces labelled with red dots and then the green dots. The lateral moving stops when all the seismic traces in the x direction are inverted.

For each group, traces of the two ending points must be included, to satisfy the periodic condition of the Fourier series.

For a given the initial model in discrete form, we set $(L_{\min}, N_{\min})$ and calculate the Fourier coefficients by fitting this initial impedance model. This step itself is an inverse problem.

In the impedance inversion, we update the Fourier coefficients iteratively, by inverting each group of seismic traces, and repeat the procedure until all trace groups are inverted.

Then, we reset (L, N) with the increments $(\Delta L, \Delta N)$. We repeat the inversion the in the last paragraph, until reaching $(L_{\max}, N_{\max})$.

We demonstrate this inversion flow using the Marmousi impedance model presented in depth-time domain (Figure 14.4a). The model size in the depth-time domain is 180×174 cells, with the distance grid interval of 12.5 m and the depth-time grid interval of 6 ms. Synthetic seismic data are shown in Figure 14.4b.

The starting impedance model is displayed in the depth-time domain, as shown in Figure 14.4c. The inversion starts from $L_{\min} = N_{\min} = 5$ with increment $\Delta L = \Delta N = 10$, and stops at $L_{\max} = N_{\max} = 55$. The maximum iteration for each set of (L, N) is 5.

We compare three inversion results using the following three schemes:

1) Perform the inversion trace by trace;

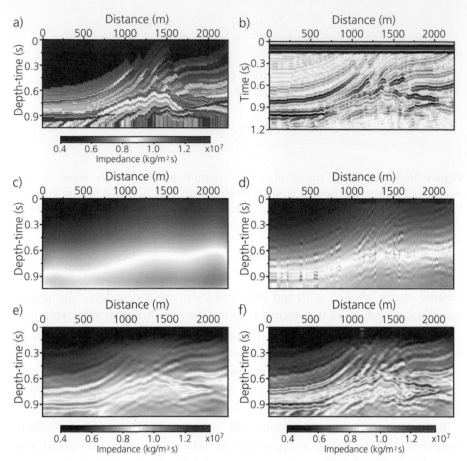

Figure 14.4 (*a*) The Marmousi impedance model presented in the depth-time domain. (*b*) The corresponding seismic profile. (*c*) The starting model for the impedance inversion. (*d*) The inverted 2-D impedance model, when using a trace by trace inversion scheme. (*e*) The inverted 2-D impedance model, when inverting all traces simultaneously. (*f*) The inverted 2-D impedance model, when grouping seismic traces along the *x* direction into four groups.

2) Perform the inversion using all of the traces simultaneously;

3) Partition the seismic section into small groups and perform the inversion using one small group at a time.

The depth-time domain impedance results are shown in Figure 14.4d–f. When using the trace-by-trace inversion scheme, the inverted model shows an obvious lateral discontinuity (Figure 14.4d). When using all of the traces

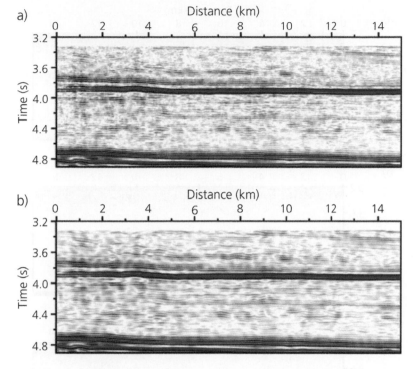

Figure 14.5 (*a*) A field seismic profile. (*b*) The synthetic seismic profile, generated from the impedance inversion.

simultaneously, the inversion model is spatially smooth (Figure 14.4e). When dividing the 2-D data into four groups along the *x* direction, but with the same inversion parameters as those used in Figure 14.4d, the reconstructed model (Figure 14.4f) has higher resolution than the model in Figure 14.4e.

The reason is that, if all the traces simultaneously make contributions to the update of the Fourier coefficients, when making a minimisation in the data misfit, the minimisation for each trace might not always be optimal. However, if the seismic traces are divided into several groups, there are two advantages. First, when the update of Fourier coefficients for the current group starts from the result of the previous group, it means that the inversion starts from a better starting model. Second, minimising the misfit of seismic traces within a small group is more likely to achieve an optimal minimisation.

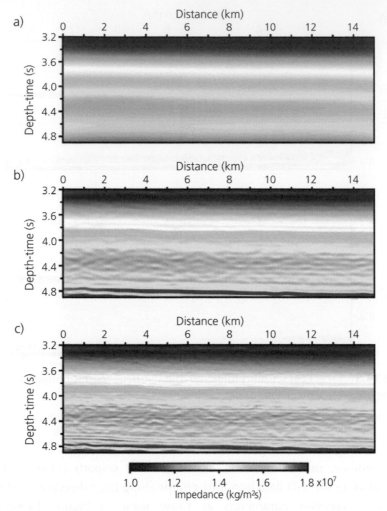

Figure 14.6 (*a*) The starting model for the impedance inversion. (*b*) Inverted impedance model, when using $L_{max} = N_{max} = 42$ in the inversion. (*c*) Inverted impedance model, when using $L_{max} = N_{max} = 92$ in the inversion.

14.4 Application to field seismic data

We test the inversion method on a 2-D field seismic profile, shown in Figure 14.5a. For a direct comparison, Figure 14.5b displays the synthetic profile, obtained after inversion (with $L_{max} = N_{max} = 92$). The synthetic profile matches the field seismic profile very well, but has much less random noise.

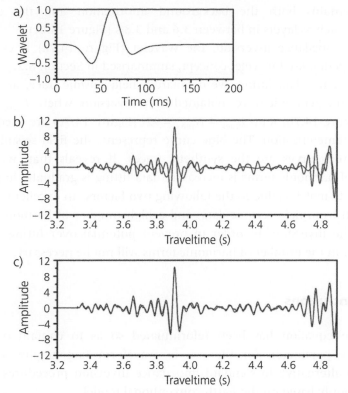

Figure 14.7 (*a*) The wavelet estimated from seismic data and used in the inversion. (*b*) The field seismic trace at $x = 7.5$ km, and the synthetic trace when using $L_{max} = N_{max} = 42$ in the inversion. (*c*) The same field seismic trace and the synthetic trace when using $L_{max} = N_{max} = 92$ in the inversion. The blue curve represents the field seismic trace and the red curves are the synthetic traces.

The inversion starts from $L = N = 2$ with the increments $\Delta L = \Delta N = 10$ along two directions. The maximum number of iterations for each (L, N) setting is 5. During the inversion, all the seismic traces along the x-direction are divided into three groups, and each group contains 200 seismic traces.

Figure 14.6a is the starting model for the inversion. Figures 14.6b and 14.6c are the inverted models, when using $L_{max} = N_{max} = 42$ and $L_{max} = N_{max} = 92$ in model parameterisation. When (L_{max}, N_{max}) is small, the background model is well recovered (Figure 14.6b). When (L_{max}, N_{max}) increases, Fourier coefficients for both low wavenumber components and high wavenumber components are updated. The inverted impedance model contains both the background information and the detailed variation, such as layers in between 3.6 and 3.8 s (Figure 14.6c).

In the impedance inversion, the wavelet (Figure 14.7a) is estimated using the generalised wavelet concept, summarised in Section 8.2.

To check the data fitting, we compare a field seismic trace, at $x = 7.5$ km, with the synthetic traces obtained from inversion, when $L_{max} = N_{max} = 42$ (Figure 14.7b) and $L_{max} = N_{max} = 92$ (Figure 14.7c) are used in the model parameterisation. The blue curve represents the field seismic trace, and the red curves are the synthetic traces. It reveals that, when the number of harmonic terms is increased, data fitting is gradually improved. A good data match is due to the following two factors: an excellent wavelet estimation from field seismic data, and a high number of harmonic terms used in the inversion. In order to prevent potential over fitting, further increment in the number of harmonic terms will not be necessary.

14.5 Conclusions

The wave equation has been reformulated so as to contain only the impedance parameter, for the use in waveform tomography. It is an advanced alternative to standard impedance inversion procedures, which are commonly based on the Earth conventional model.

The subsurface impedance model is parameterised by a truncated Fourier series, and the aim of the inversion is to extract the coefficients of various harmonic terms. This model parameterisation scheme takes seismic information from the entire study region, and guarantees a spatial continuity in the inverted model. It is a multi-scale inversion, proceeding from low wavenumbers to high wavenumbers. It inverts first for the Fourier coefficients of low wave-number components, which represents the background of the model, and then for the coefficients of high-wavenumber components, which represents details of the model.

For the inversion of field seismic data, this chapter proposes a trace-grouping strategy, which divides seismic traces into several groups where each group covers the entire study region. This strategy leads to an inversion model with a good spatial continuity and a high resolution. Further developments shall be to derive two- and three-dimensional wave equations which contain only the impedance parameter, and to perform waveform tomography in the pre-stack domain, using these single elastic-parameter equations, for the impedance inversion.

For both the Earth convolutional model-based and the wave equation-based seismic inversion methods, we have terminated the discussion at the impedance inversion. This is because the impedance variation is the direct information recorded in seismic data. While reconstructed impedance images can be used straightforwardly for reservoir characterisation, they can also be inverted further to extract various physical parameters and petro-physical parameters, for refining reservoir characterisation and hydrocarbon detection.

Appendices

Appendix A: Householder transform for QR decomposition

For a full rank $M \times N$ matrix $\mathbf{G}$, $M \geq N$, QR decomposition may transfer it to be the multiplication of an $M \times M$ orthogonal matrix $\mathbf{Q}$ and an $N \times N$ upper triangular matrix $\mathbf{R}$:

$$\mathbf{G} = \mathbf{Q} \begin{bmatrix} \mathbf{R} \\ \mathbf{0} \end{bmatrix}, \tag{A.1}$$

where $\mathbf{Q}^{\mathrm{T}}\mathbf{Q} = \mathbf{I}_{M \times M}$, and $\mathbf{0}$ is an $(M - N) \times N$ null matrix. It can be calculated by Householder transformation, which takes a column vector $\mathbf{r}$ of matrix $\mathbf{G}$ and reflects it about some plane or hyper-plane (Householder, 1955, 1964).

For a vector $\mathbf{r}$, an orthogonal matrix $\mathbf{P}_j$ can make an element of $\mathbf{r}$ to be 0:

$$
\begin{array}{ccc}
\mathbf{P}_j & \mathbf{r} & = \mathbf{P}_j \mathbf{r}
\end{array}
$$

$$
\begin{bmatrix}
1 & & & & & & & & \\
 & \ddots & & & & & & & \\
 & & 1 & & & & & & \\
 & & & c & & s & & & \\
 & & & & 1 & & & & \\
 & & & & & \ddots & & & \\
 & & & & & & 1 & & \\
 & & & -s & & c & & & \\
 & & & & & & & 1 & \\
 & & & & & & & & \ddots \\
 & & & & & & & & & 1
\end{bmatrix}
\begin{bmatrix}
\times \\ \vdots \\ \times \\ \xi_1 \\ \times \\ \vdots \\ \times \\ \xi_2 \\ \times \\ \vdots \\ \times
\end{bmatrix}
=
\begin{bmatrix}
\times \\ \vdots \\ \times \\ \rho \\ \times \\ \vdots \\ \times \\ 0 \\ \times \\ \vdots \\ \times
\end{bmatrix}, \tag{A.2}
$$

where $\times$ denotes any nonzero element in the vector $\mathbf{r}$, and

$$\rho = \sqrt{\xi_1^2 + \xi_2^2}, \qquad c = \frac{\xi_1}{\rho}, \qquad s = \frac{\xi_2}{\rho}. \tag{A.3}$$

This solution reveals that $c = \cos\theta$ and $s = \sin\theta$. Hence, the orthogonal matrix is a rotation matrix with angle θ.

Seismic Inversion: Theory and Applications, First Edition. Yanghua Wang.
© 2017 John Wiley & Sons, Ltd. Published 2017 by John Wiley & Sons, Ltd.

The orthogonal matrix $\mathbf{P}_j$ makes the jth element of $\mathbf{r}$ to be 0; k of such orthogonal matrices will make k elements be 0, and there are only $M - k$ nonzero elements left in vector $\mathbf{r}$. For an $M \times N$ matrix $\mathbf{G}$, we can implement this rotation process column by column, so that $\mathbf{G}$ becomes an upper triangular matrix:

$$\mathbf{Q}^\mathrm{T} \mathbf{G} = \begin{bmatrix} \mathbf{R} \\ \mathbf{0} \end{bmatrix}, \tag{A.4}$$

where $\mathbf{Q}^\mathrm{T}$ is an $M \times M$ matrix given by $\mathbf{Q}^\mathrm{T} = \mathbf{P}_\ell \cdots \mathbf{P}_2 \mathbf{P}_1$, and $\ell = \frac{1}{2} N \times (2M - N - 1)$. Since $\mathbf{Q}^\mathrm{T}$ is formed by the multiplication of orthogonal matrices, it is also an orthogonal matrix.

For speeding up QR decomposition, the Householder transformation can make vector $\mathbf{r}$ directly to be a vector that has only one nonzero element:

$$\mathbf{Pr} = \mathbf{e}_1, \tag{A.5}$$

where $\mathbf{e}_1 = (\rho, 0, 0, \cdots, 0)^\mathrm{T}$, and $\mathbf{P}$ is an orthogonal matrix to be determined.

The Householder algorithm is summarised as

$$\rho = -\operatorname{sign}(\xi_1) \| \mathbf{r} \|,$$
$$\mathbf{u} = \mathbf{r} - \mathbf{e}_1, \tag{A.6}$$
$$\mathbf{P} = \mathbf{I} - 2 \frac{\mathbf{u} \mathbf{u}^\mathrm{T}}{\| \mathbf{u} \|^2},$$

where ξ_1 is the first element of $\mathbf{r}$.

In this algorithm, the first equation shows that $\| \mathbf{r} \|^2 = \rho^2 = \| \mathbf{e}_1 \|^2$. It means that the orthogonal transformation does not change the norm of vector $\mathbf{r}$.

From the second equation, we know that the first element of $\mathbf{u}$ is $u_1 = \xi_1 - \rho$. Then we have

$$\| \mathbf{u} \|^2 = \| \mathbf{r} - \mathbf{e}_1 \|^2 = \mathbf{r}^\mathrm{T} \mathbf{r} - 2 \mathbf{r}^\mathrm{T} \mathbf{e}_1 + \mathbf{e}_1^\mathrm{T} \mathbf{e}_1 = \rho^2 - 2 \rho \xi_1 + \rho^2$$
$$= -2 \rho u_1, \tag{A.7}$$

which leads to

$$\mathbf{P} = \mathbf{I} + \frac{1}{\rho u_1} \mathbf{u} \mathbf{u}^\mathrm{T}. \tag{A.8}$$

Now, we may verify that

$$\mathbf{Pr} = \mathbf{r} + \frac{\mathbf{u}^\mathrm{T} \mathbf{r}}{\rho u_1} \mathbf{u} = \mathbf{r} + \frac{\mathbf{r}^\mathrm{T} \mathbf{r} - \mathbf{e}_1^\mathrm{T} \mathbf{r}}{\rho u_1} \mathbf{u} = \mathbf{r} + \frac{\rho(\rho - \xi_1)}{\rho u_1} \mathbf{u} = \mathbf{r} - \mathbf{u} \tag{A.9}$$
$$= \mathbf{e}_1.$$

Hence, $\mathbf{P}$ is the Householder matrix that can be used to reflect a vector in such a way that all coordinates but one disappear, and, hence, the Householder transformation is also called Householder reflection.

The third equation in (A.6) defines the orthogonal matrix $\mathbf{P}$. We may verify its orthogonality as

$$\mathbf{P}^{\mathrm{T}}\mathbf{P} = \left(\mathbf{I} - 2\frac{\mathbf{u}\mathbf{u}^{\mathrm{T}}}{\|\mathbf{u}\|^2}\right)^{\mathrm{T}}\left(\mathbf{I} - 2\frac{\mathbf{u}\mathbf{u}^{\mathrm{T}}}{\|\mathbf{u}\|^2}\right)$$

$$= \mathbf{I} - 4\frac{\mathbf{u}\mathbf{u}^{\mathrm{T}}}{\|\mathbf{u}\|^2} + 4\frac{\mathbf{u}(\mathbf{u}^{\mathrm{T}}\mathbf{u})\mathbf{u}^{\mathrm{T}}}{\|\mathbf{u}\|^4} \tag{A.10}$$

$$= \mathbf{I}.$$

When we choose the first matrix column for $\mathbf{r}$, we obtain the Householder matrix $\mathbf{P}_1$. Multiplying $\mathbf{G}$ with the Householder matrix $\mathbf{P}_1$ results in a matrix $\mathbf{P}_1\mathbf{G}$ with zeros in the left column (except for the first row).

$$\mathbf{P}_1\mathbf{G} = \begin{bmatrix} \rho_1 & \times & \times & \times \\ 0 & & & \\ \vdots & & \mathbf{G}' & \\ 0 & & & \end{bmatrix}. \tag{A.11}$$

This can be repeated for $\mathbf{G}'$, obtained from $\mathbf{P}_1\mathbf{G}$ by deleting the first row and first column, resulting in a Householder matrix $\mathbf{P}_2'$. Note that $\mathbf{P}_2'$ is smaller than $\mathbf{P}_1$. Since we want it really to operate on $\mathbf{P}_1\mathbf{G}$ instead of $\mathbf{G}'$ we need to expand it to the upper left, filling in a 1,

$$\mathbf{P}_2 = \begin{bmatrix} 1 & 0 & \cdots & 0 \\ 0 & & & \\ \vdots & & \mathbf{P}_2' & \\ 0 & & & \end{bmatrix}, \tag{A.12}$$

or in general,

$$\mathbf{P}_k = \begin{bmatrix} \mathbf{I}_{k-1} & 0 & \cdots & 0 \\ 0 & & & \\ \vdots & & \mathbf{P}_k' & \\ 0 & & & \end{bmatrix}. \tag{A.13}$$

The $M \times N$ matrix $\mathbf{G}$ is gradually transformed to an upper triangular matrix:

$$\mathbf{P}_\ell \cdots \mathbf{P}_2\mathbf{P}_1\mathbf{G} = \begin{bmatrix} \mathbf{R} \\ \mathbf{0} \end{bmatrix}, \tag{A.14}$$

where ℓ is the total number of iterations of this process, $\ell = \min(M-1, N)$. Meanwhile,

$$\mathbf{Q} = \mathbf{P}_1^{\mathrm{T}}\mathbf{P}_2^{\mathrm{T}} \cdots \mathbf{P}_\ell^{\mathrm{T}}, \tag{A.15}$$

defines the orthogonal matrix for the QR decomposition.

Appendix B: Singular value decomposition algorithm

This appendix summarises a practically stable algorithm for singular value decomposition (SVD), which is a two-phase, transformation-based method.

B.1   Bidiagonalisation

The first phase of SVD is reducing the matrix to a bidiagonal form. For bidiagonalisation, we can use the following two methods.

The first method is the Householder transformation, as used in QR decomposition, to eliminate elements in column vectors. After the first column transformation, only the diagonal element in the first column is nonzero. After the second column transformation, all elements except of the first and the second ones are 0. Doing for all columns, we have an upper triangular matrix:

$$\mathbf{U}_N^T \cdots \mathbf{U}_2^T \mathbf{U}_1^T \mathbf{G} = \begin{bmatrix} \otimes & \times & \times & \times & \times \\ 0 & \otimes & \times & \times & \times \\ 0 & 0 & \otimes & \times & \times \\ 0 & 0 & 0 & \otimes & \times \\ 0 & 0 & 0 & 0 & \otimes \\ 0 & 0 & 0 & 0 & 0 \end{bmatrix}, \tag{B.1}$$

where $\mathbf{U}_k^T$ is the transformation operator for the kth column, and $k = 1, 2, \cdots, N$.

Then, we can perform the Householder transformation along row vectors. Note that the elements in main diagonal ($\otimes$) have been changed during the column transformation. As they are related to the column transformation, these elements cannot be further altered. Thus, the row transformation now only changes the elements along the second diagonal. After the first row transformation, all elements except for the first and the second elements ($\otimes$) are 0. After the second row transformation, all elements except for the second and third elements ($\otimes$) are 0. Finally, the matrix is in an upper bidiagonal form:

$$\mathbf{U}_N^T \cdots \mathbf{U}_2^T \mathbf{U}_1^T \mathbf{G} \mathbf{V}_1 \cdots \mathbf{V}_{N-2} = \begin{bmatrix} \otimes & \otimes & 0 & 0 & 0 \\ 0 & \otimes & \otimes & 0 & 0 \\ 0 & 0 & \otimes & \otimes & 0 \\ 0 & 0 & 0 & \otimes & \otimes \\ 0 & 0 & 0 & 0 & \otimes \\ 0 & 0 & 0 & 0 & 0 \end{bmatrix}. \tag{B.2}$$

Denoting $\mathbf{U}_N^T \cdots \mathbf{U}_2^T \mathbf{U}_1^T = \mathbf{U}_B^T$, $\mathbf{V}_1 \cdots \mathbf{V}_{N-2} = \mathbf{V}_B$, and the upper bidiagonal matrix by $\mathbf{B}$, we have

$$\mathbf{U}_B^T \mathbf{G} \mathbf{V}_B = \mathbf{B} . \tag{B.3}$$

An alternative method is to perform bidiagonalisation directly. If **B** is the bidiagonal matrix, according to Equation B.3, we have the formula

$$\mathbf{G} = \mathbf{U}_\mathrm{B} \mathbf{B} \mathbf{V}_\mathrm{B}^\mathrm{T}.$$ (B.4)

This is related to the Lanczos algorithm. Post-multiplying **G** by $\mathbf{V}_\mathrm{B}$ and pre-multiplying **G** by $\mathbf{U}_\mathrm{B}^\mathrm{T}$ produce two equations:

$$\mathbf{G}\mathbf{V}_\mathrm{B} = \mathbf{U}_\mathrm{B}\mathbf{B}, \qquad \mathbf{G}^\mathrm{T}\mathbf{U}_\mathrm{B} = \mathbf{V}_\mathrm{B}\mathbf{B}^\mathrm{T}.$$ (B.5)

We may present these two equations explicitly as $(M > N)$

$$
\begin{bmatrix} & & \\ & \mathbf{G} & \\ & & \end{bmatrix}
\begin{bmatrix} \mathbf{v}_1 & \mathbf{v}_2 & \cdots & \mathbf{v}_N \end{bmatrix}
$$

$$
= \begin{bmatrix} \mathbf{u}_1 & \mathbf{u}_2 & \cdots & \mathbf{u}_N & \cdots & \mathbf{u}_M \end{bmatrix}
\begin{bmatrix}
\alpha_1 & \beta_1 & & & \\
& \alpha_2 & \beta_2 & & \\
& & \ddots & \ddots & \\
& & & \alpha_{N-1} & \beta_{N-1} \\
& & & & \alpha_N
\end{bmatrix}, \quad (B.6)
$$

and

$$
\begin{bmatrix} & & \\ & \mathbf{G}^\mathrm{T} & \\ & & \end{bmatrix}
\begin{bmatrix} \mathbf{u}_1 & \mathbf{u}_2 & \cdots & \mathbf{u}_N & \cdots & \mathbf{u}_M \end{bmatrix}
$$

$$
= \begin{bmatrix} \mathbf{v}_1 & \mathbf{v}_2 & \cdots & \mathbf{v}_N \end{bmatrix}
\begin{bmatrix}
\alpha_1 & & & & \\
\beta_1 & \alpha_2 & & & \\
& \beta_2 & \ddots & & \\
& & \ddots & \alpha_{N-1} & \\
& & & \beta_{N-1} & \alpha_N
\end{bmatrix}. \quad (B.7)
$$

Based on this diagrammatic illustration, we can generate vectors $[\mathbf{u}_1, \mathbf{u}_2, \cdots, \mathbf{u}_N]$ and $[\mathbf{v}_1, \mathbf{v}_2, \cdots, \mathbf{v}_N]$ recursively from the following two equations:

$$\alpha_k \mathbf{u}_k = \mathbf{G}\mathbf{v}_k - \beta_{k-1}\mathbf{u}_{k-1},$$

$$\beta_k \mathbf{v}_{k+1} = \mathbf{G}^\mathrm{T}\mathbf{u}_k - \alpha_k \mathbf{v}_k, \tag{B.8}$$

in which α_k and β_k are both computed as normalising factors to ensure that $\mathbf{u}_k^\mathrm{T}\mathbf{u}_k = \mathbf{v}_{k+1}^\mathrm{T}\mathbf{v}_{k+1} = \mathbf{I}$. The implementation may be summarised as in what follows.

Choose the first vector $\mathbf{v}_1$ to be an arbitrary unit-norm vector, such that $\mathbf{v}_1^\mathrm{T}\mathbf{v}_1 = \mathbf{I}$. Given $\beta_0 = 0$, then calculate recursively for $k = 1, 2, \cdots, N$:

$$\mathbf{p}_k = \mathbf{G}\mathbf{v}_k - \beta_{k-1}\mathbf{u}_{k-1}, \qquad \alpha_k = \|\mathbf{p}_k\|, \qquad \mathbf{u}_k = \frac{1}{\alpha_k}\mathbf{p}_k,$$

$$\mathbf{q}_{k+1} = \mathbf{G}^\mathrm{T}\mathbf{u}_k - \alpha_k \mathbf{v}_k, \qquad \beta_k = \|\mathbf{q}_{k+1}\|, \qquad \mathbf{v}_{k+1} = \frac{1}{\beta_k}\mathbf{q}_{k+1}. \tag{B.9}$$

The other vectors $[\mathbf{u}_{N+1}, \cdots, \mathbf{u}_M]$ are of no relevance to the decomposition. This algorithm is referred to as Golub-Kahan-Lanczos bidiagonalisation (Golub and Kahan, 1965).

B.2 Diagonalisation

The second phase is SVD of an upper bidiagonal matrix. It may be achieved by a variant of the QR method. The simplest method is the Jacobi diagonalisation as in what follows.

As the singular values of the upper bidiagonal matrix $\mathbf{B}$ is the square root of the eigenvalues of the following matrix:

$$\mathbf{T} = \mathbf{B}^\mathrm{T}\mathbf{B}. \tag{B.10}$$

This is a symmetric tridiagonal matrix:

$$\mathbf{T} = \begin{bmatrix} a_1 & b_1 & & & & \\ b_1 & a_2 & b_2 & & & \\ & b_2 & a_3 & b_3 & & \\ & & \ddots & \ddots & \ddots & \\ & & & b_{N-2} & a_{N-1} & b_{N-1} \\ & & & & b_{N-1} & a_N \end{bmatrix}. \tag{B.11}$$

Each Jacobi transformation eliminates one pair of off-diagonal elements in this symmetric matrix. In order to eliminate the pair of equal elements b_i, Jacobi method employs an orthogonal transformation matrix such as

$$\text{col } i$$

$$
\mathbf{P} = \begin{bmatrix}
\ddots & & & & \\
& 1 & & & \\
& & c & s & \\
& & -s & c & \\
& & & & 1 \\
& & & & & \ddots
\end{bmatrix} \quad \text{row } i \qquad (B.12)
$$

Transformation $\mathbf{PTP}^{\mathrm{T}}$ only affects the row i and $i+1$ and columns i and $i+1$:

$$
\begin{bmatrix}
1 & & & \\
& c & s & \\
& -s & c & \\
& & & 1
\end{bmatrix}
\begin{bmatrix}
a_{i-1} & b_{i-1} & & \\
b_{i-1} & a_i & b_i & \\
& b_i & a_{i+1} & b_{i+1} \\
& & b_{i+1} & a_{i+2}
\end{bmatrix}
\begin{bmatrix}
1 & & & \\
& c & -s & \\
& s & c & \\
& & & 1
\end{bmatrix}
$$

$$
= \begin{bmatrix}
a_{i-1} & b_{i-1}c & & \\
b_{i-1}c & a_i c^2 + a_{i+1}s^2 + 2b_i cs & (-a_i + a_{i+1})cs + b_i(c^2 - s^2) & \\
& (-a_i + a_{i+1})cs + b_i(c^2 - s^2) & a_i s^2 + a_{i+1}c^2 - 2b_i cs & b_{i+1}c \\
& & b_{i+1}c & a_{i+2}
\end{bmatrix}
$$

$$
= \begin{bmatrix}
a_{i-1} & \bar{b}_{i-1} & & \\
\bar{b}_{i-1} & \bar{a}_i & \bar{b}_i & \\
& \bar{b}_i & \bar{a}_{i+1} & \bar{b}_{i+1} \\
& & \bar{b}_{i+1} & a_{i+2}
\end{bmatrix}. \qquad (B.13)
$$

That is, only two diagonal elements ($\bar{a}_i$, $\bar{a}_{i+1}$) and three off-diagonal elements ($\bar{b}_{i-1}$, $\bar{b}_i$, $\bar{b}_{i+1}$) in $\mathbf{T}^{(k+1)}$ have been modified. The transformation in Equation B.13 makes the element $\bar{b}_i = 0$. Therefore, we have a system of equations

$$
c^2 + s^2 = 1,
$$
$$
(-a_i + a_{i+1})cs + b_i(c^2 - s^2) = 0. \qquad (B.14)
$$

The solution to this system is

$$
c^2 = \frac{1}{2}\left(1 + \frac{a_i - a_{i+1}}{r}\right), \qquad (B.15a)
$$

$$
s^2 = \frac{1}{2}\left(1 - \frac{a_i - a_{i+1}}{r}\right), \qquad (B.15b)
$$

$$cs = \frac{b_i}{r}, \tag{B.15c}$$

where

$$r = \sqrt{(a_i - a_{i+1})^2 + 4b_i^2} . \tag{B.15d}$$

If $a_i > a_{i+1}$, we should find c using Equation B.15a, given the positive sign, and then find s using Equation B.15c, in which s has the same sign as b_i. Alternatively, if $a_i < a_{i+1}$, we should find s using Equation B.15b and then find c using Equation B.15c. By proceeding in this way, there is no significant loss in accuracy with floating point arithmetic, even when b_i^2 is much smaller than $(a_i - a_{i+1})^2$.

According to Equation B.13, the two diagonal elements modified by the transformation become

$$\bar{a}_i = \frac{1}{2}(a_i + a_{i+1} + r), \qquad \bar{a}_{i+1} = \frac{1}{2}(a_i + a_{i+1} - r). \tag{B.16}$$

The other two elements affected by the transformation are

$$\bar{b}_{i-1} = b_{i-1}c, \qquad \bar{b}_{i+1} = b_{i+1}c. \tag{B.17}$$

The normal procedure is to perform a series of transformations of the type described above, with each transformation eliminating the off-diagonal element having the largest modulus present in the matrix at that stage. However, elements that have been eliminated do not necessarily stay zero, and hence the method is iterative in character.

After ℓ transformations, the matrix $\mathbf{T}$ becomes a diagonal matrix $\mathbf{D}$,

$$\mathbf{P}_\ell \cdots \mathbf{P}_2 \mathbf{P}_1 \, \mathbf{T} \, \mathbf{P}_1^T \mathbf{P}_2^T \cdots \mathbf{P}_\ell^T = \mathbf{D} . \tag{B.18}$$

Denoting

$$\mathbf{Q} = \mathbf{P}_1^T \mathbf{P}_2^T \cdots \mathbf{P}_\ell^T , \tag{B.19}$$

we have

$$\mathbf{T} = \mathbf{QDQ}^T . \tag{B.20}$$

Let us now revisit the original upper bidiagonal matrix $\mathbf{B}$. If SVD of $\mathbf{B}$ is in the form of $\mathbf{B} = \mathbf{U}_D \Lambda \mathbf{V}_D^T$, then

$$\mathbf{B}^T\mathbf{B} = [\mathbf{U}_D \Lambda \mathbf{V}_D^T]^T \mathbf{U}_D \Lambda \mathbf{V}_D^T = \mathbf{V}_D \Lambda^2 \mathbf{V}_D^T . \tag{B.21}$$

Comparing Equation B.20 and Equation B.21, we obtain the singular-value matrix $\Lambda = \mathbf{D}^{1/2}$ and the right eigenvector matrix $\mathbf{V}_D = \mathbf{Q}$. We can also calculate the left eigenvector matrix $\mathbf{U}_D$ by

$$\mathbf{U}_D = \mathbf{B}\mathbf{V}_D \Lambda^{-1} . \tag{B.22}$$

Combining two steps, $\mathbf{G} = \mathbf{U}_\mathrm{B} \mathbf{B} \mathbf{V}_\mathrm{B}^\mathrm{T}$, and $\mathbf{B} = \mathbf{U}_\mathrm{D} \boldsymbol{\Lambda} \mathbf{V}_\mathrm{D}^\mathrm{T}$ we have the final SVD of $\mathbf{G}$, as

$$\mathbf{G} = \mathbf{U}_\mathrm{B} \mathbf{U}_\mathrm{D} \, \mathbf{D}^{1/2} \, \mathbf{V}_\mathrm{D}^\mathrm{T} \mathbf{V}_\mathrm{B}^\mathrm{T} = \mathbf{U} \boldsymbol{\Lambda} \mathbf{V}^\mathrm{T} . \tag{B.23}$$

Note that the left eigenvector matrix is updated $\mathbf{U} = \mathbf{U}_\mathrm{B} \mathbf{U}_\mathrm{D}$, and the right eigenvector matrix is also updated $\mathbf{V} = \mathbf{V}_\mathrm{B} \mathbf{V}_\mathrm{D}$.

Appendix C: Biconjugate gradient method for complex systems

For solving the linear system $\mathbf{G}\mathbf{x} = \mathbf{d}$, Lanczos (1950) and then Fletcher (1976) proposed a biconjugate gradient method, in which the residual $\mathbf{e}^{(k)}$ is orthogonal with respect to another row of vectors $\hat{\mathbf{e}}^{(1)}$, $\hat{\mathbf{e}}^{(2)}$, $\cdots$, $\hat{\mathbf{e}}^{(k)}$, and, vice versa, $\hat{\mathbf{e}}^{(k)}$ is orthogonal with respect to the row of vectors $\mathbf{e}^{(1)}$, $\mathbf{e}^{(2)}$, $\cdots$, $\mathbf{e}^{(k)}$. This is the biorthogonality condition:

$$(\hat{\mathbf{e}}^{(k)}, \mathbf{e}^{(i)}) = (\mathbf{e}^{(k)}, \hat{\mathbf{e}}^{(i)}) = 0, \quad \text{for all } i < k. \tag{C.1}$$

In addition, two rows of search directions $\mathbf{p}^{(k)}$ and $\hat{\mathbf{p}}^{(k)}$ also satisfy the biconjugacy condition:

$$(\hat{\mathbf{p}}^{(k)}, \mathbf{G}\mathbf{p}^{(i)}) = (\mathbf{p}^{(k)}, \mathbf{G}\hat{\mathbf{p}}^{(i)}) = 0, \quad \text{for all } i < k. \tag{C.2}$$

Many seismic inverse problems, such as the frequency-domain waveform tomography presented in Chapter 12, have a complex-valued system of linear equations. Here is the summary of a biconjugate gradient method, for solving a non-symmetrical complex system.

For complex matrix $\mathbf{G}$ and vector $\mathbf{d}$, given an initial solution $\mathbf{x}^{(1)}$, computing $\mathbf{e}^{(1)} = \mathbf{d} - \mathbf{G}\mathbf{x}^{(1)}$; Then, set $\hat{\mathbf{e}}^{(1)} = \overline{\mathbf{e}}^{(1)}$, where $\overline{\mathbf{e}}$ is the complex-conjugate vector of $\mathbf{e}$; also, set $\mathbf{p}^{(1)} = \mathbf{e}^{(1)}$, and $\hat{\mathbf{p}}^{(1)} = \overline{\mathbf{p}}^{(1)}$.

Iterations for $k = 1, 2, \cdots$:

1) Calculatie the step length,

$$\alpha_k = \frac{(\hat{\mathbf{e}}^{(k)}, \mathbf{e}^{(k)})}{(\hat{\mathbf{p}}^{(k)}, \mathbf{G}\mathbf{p}^{(k)})} , \tag{C.3}$$

for updating the solution, $\mathbf{x}^{(k+1)} = \mathbf{x}^{(k)} + \alpha_k \mathbf{p}^{(k)}$. Correspondingly, the residual and the biresidual are

$$\mathbf{e}^{(k+1)} = \mathbf{e}^{(k)} - \alpha_k \mathbf{G}\mathbf{p}^{(k)}, \tag{C.4}$$

$$\hat{\mathbf{e}}^{(k+1)} = \hat{\mathbf{e}}^{(k)} - \overline{\alpha}_k \mathbf{G}^\mathrm{H} \hat{\mathbf{p}}^{(k)}, \tag{C.5}$$

where $\overline{\alpha}_k$ is the conjugate of complex number α_k, and superscript $^\mathrm{H}$ represents Hermitian transpose of the complex conjugate. For two vectors $\mathbf{u}$ and $\mathbf{v}$, $(\mathbf{u}, \mathbf{v}) = \mathbf{u}^\mathrm{H}\mathbf{v}$, and $(\mathbf{u}, \mathbf{G}\mathbf{v}) = (\mathbf{G}^\mathrm{H}\mathbf{u}, \mathbf{v})$.

2) Define the biconjugate coefficient

$$\beta_k = \frac{(\hat{\mathbf{e}}^{(k+1)}, \mathbf{e}^{(k+1)})}{(\hat{\mathbf{e}}^{(k)}, \mathbf{e}^{(k)})} . \tag{C.6}$$

Then, the search direction $\mathbf{p}^{(k+1)}$ and bidirection $\hat{\mathbf{p}}^{(k+1)}$ are

$$\mathbf{p}^{(k+1)} = \mathbf{e}^{(k+1)} + \beta_k \mathbf{p}^{(k)} , \tag{C.7}$$

$$\hat{\mathbf{p}}^{(k+1)} = \hat{\mathbf{r}}^{(k+1)} + \overline{\beta}_k \hat{\mathbf{p}}^{(k)} , \tag{C.8}$$

respectively, where $\overline{\beta}_k$ is the conjugate of complex number β_k.

Note that, after the first iteration, $k \ge 2$, the biresidual $\hat{\mathbf{e}}^{(k)}$ is *not* usually the complex conjugate of the residual $\mathbf{e}^{(k)}$, and in turn the bidirection $\hat{\mathbf{p}}^{(k)}$ is *not* usually the complex conjugate of the direction $\mathbf{p}^{(k)}$.

At the end of the first step, if $\| \mathbf{e}^{(k+1)} \| < \varepsilon \| \mathbf{b} \|$, in which ε is a small positively valued threshold, $\| \cdot \|$ is the L_2 norm of a vector, then the output is the solution $\mathbf{x}^{(k+1)}$; Otherwise continuing to the second step.

Appendix D: Gradient calculation in waveform tomography

For seismic waveform tomography, let us set an objective function as

$$\phi(\mathbf{m}) = \frac{1}{2} \| \mathbf{u}_{\text{obs}} - \mathbf{u}(\mathbf{m}) \|^2 , \tag{D.1}$$

where $\mathbf{u}_{\text{obs}}$ is the observed seismic trace, and $\mathbf{u}(\mathbf{m})$ is an synthetic seismic trace, calculated based on model $\mathbf{m}$. Here, we deal with a time domain waveform tomography, using the 1-D acoustic wave equation, presented in Chapter 14:

$$4 \frac{\partial}{\partial \tau} \left(Z(\tau) \frac{\partial u(\tau, t)}{\partial \tau} \right) - Z(\tau) \frac{\partial^2 u(\tau, t)}{\partial t^2} = -\delta(\tau - \tau_s) s(t) , \tag{D.2}$$

where τ is the depth-time (with the units of time), $Z(\tau)$ is the impedance parameter, t is wave travelling time, $u(\tau, t)$ is the wavefield of displacement, $s(t)$ is the source function, and τ_s denotes the source position. Hence, the model vector $\mathbf{m} \equiv Z(\tau)$ is the impedance function varying along depth-time τ.

This appendix derives the gradient vector of the objective function in Equation D.1, with respect to the impedance $Z(\tau)$:

$$\frac{\partial \phi}{\partial Z} = -\left[\frac{\partial \mathbf{u}}{\partial Z} \right]^{\text{T}} \Delta \mathbf{u} . \tag{D.3}$$

Following the theory of seismic scattering (section 2.2), any perturbation in the impedance may cause a perturbation in the wavefield:

$$Z(\tau) = Z_0(\tau) + \delta Z(\tau),$$

$$u(\tau, t; \tau_s) = u_0(\tau, t; \tau_s) + \delta u(\tau, t; \tau_s). \tag{D.4}$$

Wavefield perturbation $\delta u(\tau, t; \tau_s)$ follows the same wave equation as Equation D.2, which can be expressed as

$$4 \frac{\partial}{\partial \tau} \left(Z_0(\tau) \frac{\partial \delta u(\tau, t; \tau_s)}{\partial \tau} \right) - Z_0(\tau) \frac{\partial^2 \delta u(\tau, t; \tau_s)}{\partial t^2} = -\Delta s(\tau, t; \tau_s), \tag{D.5}$$

where $\Delta s(\tau, t; \tau_s)$ is the virtual source,

$$\Delta s(\tau, t; \tau_s) \approx 4 \frac{\partial}{\partial \tau} \left(\delta Z(\tau) \frac{\partial u_0(\tau, t; \tau_s)}{\partial \tau} \right) - \delta Z(\tau) \frac{\partial^2 u_0(\tau, t; \tau_s)}{\partial t^2}. \tag{D.6}$$

An analytical solution of Equation D.5 can be expressed in terms of Green's function, $g(\tau, t; \tau', t')$, as

$$\delta u(\tau, t; \tau_s) = \int g(\tau, t; \tau', 0) * \Delta s(\tau', t; \tau_s) d\tau', \tag{D.7}$$

where $*$ denotes a convolution along the time variable t. Using the reciprocity theorem for the Green's function, we can rewrite (D.7) as

$$\delta u(\tau, t; \tau_s) = \int g(\tau', t; \tau, 0) * \Delta s(\tau', t; \tau_s) d\tau'. \tag{D.8}$$

Substituting Equation D.6 into Equation D.8, we have

$$\delta u(\tau, t; \tau_s) = \int \left[4 g(\tau', t; \tau, 0) * \frac{\partial}{\partial \tau'} \left(\delta Z(\tau') \frac{\partial u_0(\tau', t; \tau_s)}{\partial \tau'} \right) \right.$$
$$\left. - g(\tau', t; \tau, 0) * \left(\delta Z(\tau') \frac{\partial^2 u_0(\tau', t; \tau_s)}{\partial t^2} \right) \right] d\tau'. \tag{D.9}$$

Integrating Equation D.9 by parts, we obtain the wavefield perturbation as

$$\delta u(\tau, t; \tau_s) = \int \left[-4 \frac{\partial g(\tau', t; \tau, 0)}{\partial \tau'} * \left(\delta Z(\tau') \frac{\partial u_0(\tau', t; \tau_s)}{\partial \tau'} \right) \right.$$
$$\left. + \frac{\partial g(\tau', t; \tau, 0)}{\partial t} * \left(\delta Z(\tau') \frac{\partial u_0(\tau', t; \tau_s)}{\partial t} \right) \right] d\tau'. \tag{D.10}$$

Here, we assume that $u_0(\tau, t; \tau_s)$ satisfies the following conditions:

$$\frac{\partial u_0(\tau = 0, t; \tau_s)}{\partial \tau} = \frac{\partial u_0(\tau \to \infty, t; \tau_s)}{\partial \tau} = 0, \qquad \frac{\partial u_0(\tau, t = 0; \tau_s)}{\partial t} = \frac{\partial u_0(\tau, t \to \infty; \tau_s)}{\partial t} = 0. \tag{D.11}$$

Utilization of Equation D.10 to solve the forward problem is known as the Born approximation.

Based on Equation D.10, the wavefield perturbation, we can now derive the Fréchet derivative of the wavefield, with respect to the impedance. The forward problem can be expressed as

$$\delta u(\tau, t; \tau_s) \approx \int \frac{\partial u(\tau, t; \tau_s \mid Z(\tau'))}{\partial Z(\tau')} \delta Z(\tau') \mathrm{d}\tau' . \qquad (D.12)$$

Comparing it to Equation D.10, and changing $\tau \rightarrow \tau_g$, $\tau' \rightarrow \tau$, we obtain the Fréchet kernel as

$$\frac{\partial u(\tau_g, t; \tau_s \mid Z(\tau))}{\partial Z(\tau)} = -4 \frac{\partial g(\tau, t; \tau_g, 0)}{\partial \tau} * \frac{\partial u_0(\tau, t; \tau_s)}{\partial \tau} + \frac{\partial g(\tau, t; \tau_g, 0)}{\partial t} * \frac{\partial u_0(\tau, t; \tau_s)}{\partial t} . \qquad (D.13)$$

In waveform tomography, this Fréchet kernel is needed for the calculation of the gradient vector. In the inversion, the Fréchet kernel may be transposed to $\partial u(Z(\tau) \mid \tau_g, t; \tau_s) / \partial Z(\tau)$, and

$$\frac{\partial u(Z(\tau) \mid \tau, t; \tau_s)}{\partial Z(\tau)} = \frac{\partial u(\tau, t; \tau_s \mid Z(\tau))}{\partial Z(\tau)} . \qquad (D.14)$$

Then, the gradient is expressed as

$$\frac{\partial \phi}{\partial Z(\tau)} = -\sum_s \int \sum_g \frac{\partial u(Z(\tau) \mid \tau_g, t; \tau_s)}{\partial Z(\tau)} \delta u(\tau_g, t; \tau_s) \mathrm{d}t . \qquad (D.15)$$

For the case with a single seismic trace, this is Equation D.3. For the convenience, we consider at this moment this general case with a single shot-receiver pair at any arbitrary spatial position:

$$\frac{\partial \phi}{\partial Z(\tau)} = -\int \frac{\partial u(Z(\tau) \mid \tau_g, t; \tau_s)}{\partial Z(\tau)} \delta u(\tau_g, t; \tau_s) \mathrm{d}t . \qquad (D.16)$$

Using Equation D.13, the gradient can be expressed as

$$\frac{\partial \phi}{\partial Z(\tau)} = \int \left(4 \frac{\partial g(\tau, t; \tau_g, 0)}{\partial \tau} * \frac{\partial u_0(\tau, t; \tau_s)}{\partial \tau} - \frac{\partial g(\tau, t; \tau_g, 0)}{\partial t} * \frac{\partial u_0(\tau, t; \tau_s)}{\partial t} \right) \delta u(\tau_g, t; \tau_s) \, \mathrm{d}t . \qquad (D.17)$$

Using the following two expressions (Tarantola, 1984)

$$\int [g(t) * h(t)] f(t) \mathrm{d}t = \int [g(-t) * f(t)] h(t) \mathrm{d}t , \qquad (D.18)$$

and

$$g(\tau, -t; \tau', 0) = g(\tau, 0; \tau', t) , \qquad (D.19)$$

Equation D.17 becomes

$$
\frac{\partial \phi}{\partial Z(\tau)} = \int \left[4\left(\frac{\partial g(\tau,0;\tau_g,t)}{\partial \tau} * \delta u(\tau_g,t;\tau_s) \right) \frac{\partial u_0(\tau,t;\tau_s)}{\partial \tau} \right.
$$

$$
\left. - \left(\frac{\partial g(\tau,0;\tau_g,t)}{\partial t} * \delta u(\tau_g,t;\tau_s) \right) \frac{\partial u_0(\tau,t;\tau_s)}{\partial t} \right] dt .
$$

(D.20)

Defining the back-propagation wavefield as

$$
u_b(\tau,t;\tau_s) = g(\tau,0;\tau_g,t) * \delta u(\tau_g,t;\tau_s) ,
$$

(D.21)

then the gradient can be written as

$$
\frac{\partial \phi}{\partial Z(\tau)} = \int \left(4\frac{\partial u_b(\tau,t;\tau_s)}{\partial \tau} \frac{\partial u_0(\tau,t;\tau_s)}{\partial \tau} - \frac{\partial u_b(\tau,t;\tau_s)}{\partial t} \frac{\partial u_0(\tau,t;\tau_s)}{\partial t} \right) dt .
$$

(D.22)

If there are multiple receivers, the right-hand side of Equation D.21 should be a summation over receivers. If there are multiple sources, the right-hand side of Equation D.22 should be a summation over all sources.

Equation D.22 indicates that the gradient vector is evaluated by cross-correlation of (the first derivatives of) two wavefields. One of them is the forward modelling wavefield $u_0(\tau,t)$, generated with the normal seismic source. Another is the back-propagation wavefield $u_b(\tau,t)$, generated with data residual as a virtual source (Tarantola, 1984; Gauthier *et al.*, 1986). Both wavefield simulations are based on the current model estimate.

Exercises and solutions

1. For the linear system $\mathbf{Gx} = \mathbf{d}$, where

$$\mathbf{G} = \begin{bmatrix} 1 & 1 \\ 10 & 11 \end{bmatrix}, \quad \mathbf{d} = \begin{bmatrix} 11 \\ 111 \end{bmatrix},$$

the solution is $\mathbf{x} = [10 \;\; 1]^{\mathrm{T}}$. Assuming the data have small errors,

$$\tilde{\mathbf{d}} = \mathbf{d} + \mathbf{e} = \begin{bmatrix} 11 \\ 111 \end{bmatrix} + \begin{bmatrix} 0.1 \\ 0 \end{bmatrix},$$

the solution of $\mathbf{Gx} = \tilde{\mathbf{d}}$ is $\mathbf{x} = [11.1 \;\; 0]^{\mathrm{T}}$. Explain why small errors in the data vector $\tilde{\mathbf{d}}$ cause a huge difference in the solution $\mathbf{x}$.

A: For the 2×2 matrix

$$\mathbf{G} = \begin{bmatrix} 1 & 1 \\ 10 & 11 \end{bmatrix},$$

two eigenvalues are

$$\lambda_{1,2} = 6 \pm \sqrt{35}.$$

The ratio of these two eigenvalues is

$$\frac{\lambda_1}{\lambda_2} = (6 + \sqrt{35})^2 \approx 142.$$

This is the condition number of the linear system. A problem with a high condition number is ill-conditioned. For this ill-conditioned problem, small data errors, $\mathbf{e} = [0.1 \;\; 0]^{\mathrm{T}}$, can cause a large perturbation in the solution, $\Delta \mathbf{x} = [1.1 \;\; -1]^{\mathrm{T}}$. ∎

2. The norm of a vector is a measure of distance. Explain the physical meaning of L_1, L_2 and L_∞ norms. For a vector $\mathbf{x} = [9, -12]^{\mathrm{T}}$, compute $\| \mathbf{x} \|_1$, $\| \mathbf{x} \|_2$, and $\| \mathbf{x} \|_\infty$.

A: The physical meaning of the L_1 norm, $\| \mathbf{x} \|_1$, is a measure of taxicab geometry, or city block distance in a rectangular street grid.

Seismic Inversion: Theory and Applications, First Edition. Yanghua Wang.
© 2017 John Wiley & Sons, Ltd. Published 2017 by John Wiley & Sons, Ltd.

The L_2 norm, $\| \mathbf{x} \|_2$, is the ordinary straight distance from the origin to the point $\mathbf{x}$.

The L_∞ norm, $\| \mathbf{x} \|_\infty$, is the longest street that the taxi drives through.

For vector $\mathbf{x} = [9, -12]^T$: the L_1 norm, $\| \mathbf{x} \|_1 = \sum_i | x_i | = 21$; the L_2 norm, $\| \mathbf{x} \|_2 = (\sum_i x_i^2)^{1/2} = 15$; and the L_∞ norm, $\| \mathbf{x} \|_\infty = \max | x_i | = 12$.   ∎

3. For a 2×2 matrix, the determinant is

$$\det(\mathbf{A}) = \begin{vmatrix} a_{11} & a_{12} \\ a_{21} & a_{22} \end{vmatrix}.$$

Show that the absolute value of the determinant of this 2×2 matrix is the area of a parallelogram formed by two row vectors.

A: Two row vectors $\mathbf{r}_1$ and $\mathbf{r}_2$ of matrix $\mathbf{A}$ form a parallelogram:

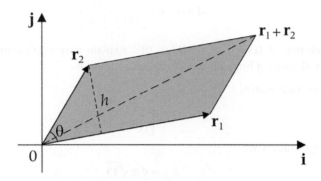

where h is the height of the parallelogram, θ is the angle between two vectors $\mathbf{r}_1$ and $\mathbf{r}_2$, and $\mathbf{i}$ and $\mathbf{j}$ are the two basis vectors.

The area of this parallelogram is the modular $\| \mathbf{r}_1 \|$ times the height $h = \| \mathbf{r}_2 \| \sin\theta$:

$$\| \mathbf{r}_1 \| h = \| \mathbf{r}_1 \| \| \mathbf{r}_2 \| \sin\theta = \| \mathbf{r}_1 \times \mathbf{r}_2 \|.$$

This equation indicates that the area of the parallelogram is the modular of the cross-product of two row vectors.

Considering these two row vectors in 3-D space, $\mathbf{r}_1 = [a_{11} \quad a_{12} \quad 0]^T$ and $\mathbf{r}_2 = [a_{21} \quad a_{22} \quad 0]^T$, the cross-product is

$$\mathbf{r}_1 \times \mathbf{r}_2 = \begin{vmatrix} \mathbf{i} & \mathbf{j} & \mathbf{k} \\ a_{11} & a_{12} & 0 \\ a_{21} & a_{22} & 0 \end{vmatrix} = \begin{vmatrix} a_{12} & 0 \\ a_{22} & 0 \end{vmatrix} \mathbf{i} - \begin{vmatrix} a_{11} & 0 \\ a_{21} & 0 \end{vmatrix} \mathbf{j} + \begin{vmatrix} a_{11} & a_{12} \\ a_{21} & a_{22} \end{vmatrix} \mathbf{k}$$

$$= 0\,\mathbf{i} - 0\,\mathbf{j} + \begin{vmatrix} a_{11} & a_{12} \\ a_{21} & a_{22} \end{vmatrix} \mathbf{k},$$

where $\mathbf{i}$, $\mathbf{j}$ and $\mathbf{k}$ are the standard basis vectors. Then, it proves that the area of the parallelogram is the absolute value of the determinant of the matrix

$$\| \mathbf{r}_1 \times \mathbf{r}_2 \| = \left| \det \begin{pmatrix} a_{11} & a_{12} \\ a_{21} & a_{22} \end{pmatrix} \right|.$$

Note that when the angle from the first to the second vector defining the parallelogram turns in a clockwise direction, the 'area' is negative valued, and so is the determinant. ∎

4. For the following 3×3 matrix,

$$\mathbf{A} = \begin{bmatrix} 2 & 1 & 1 \\ 3 & 2 & 1 \\ 2 & 1 & 2 \end{bmatrix},$$

using the Gauss-Jordan elimination method to solve the inverse matrix.

A: The augmented matrix $[\mathbf{A} \,|\, \mathbf{I}]$ is

$$\begin{bmatrix} 2 & 1 & 1 & 1 & 0 & 0 \\ 3 & 2 & 1 & 0 & 1 & 0 \\ 2 & 1 & 2 & 0 & 0 & 1 \end{bmatrix}.$$

Let us denote the kth row by $\mathbf{r}_k$, and perform the elementary row operations step by step as in the following.

$$\begin{matrix} \mathbf{r}_2 - \frac{3}{2}\mathbf{r}_1 \Rightarrow \mathbf{r}_2 \\ \mathbf{r}_3 - \mathbf{r}_1 \Rightarrow \mathbf{r}_3 \end{matrix} \quad \begin{bmatrix} 2 & 1 & 1 & 1 & 0 & 0 \\ 0 & \frac{1}{2} & -\frac{1}{2} & -\frac{3}{2} & 1 & 0 \\ 0 & 0 & 1 & -1 & 0 & 1 \end{bmatrix},$$

$$\mathbf{r}_2 + \frac{1}{2}\mathbf{r}_3 \Rightarrow \mathbf{r}_2 \quad \begin{bmatrix} 2 & 1 & 1 & 1 & 0 & 0 \\ 0 & \frac{1}{2} & 0 & -2 & 1 & \frac{1}{2} \\ 0 & 0 & 1 & -1 & 0 & 1 \end{bmatrix},$$

$$\mathbf{r}_1 - 2\mathbf{r}_2 \Rightarrow \mathbf{r}_1 \quad \begin{bmatrix} 2 & 0 & 1 & 5 & -2 & -1 \\ 0 & \frac{1}{2} & 0 & -2 & 1 & \frac{1}{2} \\ 0 & 0 & 1 & -1 & 0 & 1 \end{bmatrix},$$

$$\mathbf{r}_1 - \mathbf{r}_3 \Rightarrow \mathbf{r}_1 \quad \begin{bmatrix} 2 & 0 & 0 & 6 & -2 & -2 \\ 0 & \frac{1}{2} & 0 & -2 & 1 & \frac{1}{2} \\ 0 & 0 & 1 & -1 & 0 & 1 \end{bmatrix}.$$

The product of pivots, $2 \times \frac{1}{2} \times 1 = 1$, is the determinant of $\mathbf{A}$.

The last Gauss-Jordan step is to divide each row by its pivot, so that the new pivots are 1:

$$\begin{array}{c} \frac{1}{2}\mathbf{r}_1 \Rightarrow \mathbf{r}_1 \\ 2\mathbf{r}_2 \Rightarrow \mathbf{r}_2 \end{array} \left[\begin{array}{ccc|ccc} 1 & 0 & 0 & 3 & -1 & -1 \\ 0 & 1 & 0 & -4 & 2 & 1 \\ 0 & 0 & 1 & -1 & 0 & 1 \end{array}\right].$$

Now the augmented matrix $[\mathbf{A}\,|\,\mathbf{I}]$ has been row-reduced to $[\mathbf{I}\,|\,\mathbf{A}^{-1}]$. The second half of this six-column matrix is the inverse matrix $\mathbf{A}^{-1}$:

$$\mathbf{A}^{-1} = \begin{bmatrix} 3 & -1 & -1 \\ -4 & 2 & 1 \\ -1 & 0 & 1 \end{bmatrix}.$$

∎

5. Prove that the following matrix is a full rank matrix:

$$\mathbf{G} = \begin{bmatrix} -1 & 3 & 4 \\ 1 & -1 & 3 \\ 0 & -1 & 2 \\ 2 & -3 & 1 \end{bmatrix}.$$

A: For a rectangular matrix with $M \geq N$, if $\text{rank}(\mathbf{G}) = N$, it is a full rank matrix.

The physical meaning of the rank is the number of independent linear equations within the system. Therefore, the basic idea for determining the rank is to reduce the given matrix to an equivalent matrix whose rank can be determined by inspection.

$$\begin{bmatrix} -1 & 3 & 4 \\ 1 & -1 & 3 \\ 0 & -1 & 2 \\ 2 & -3 & 1 \end{bmatrix} \Rightarrow \begin{array}{c} \\ \mathbf{r}_1 + \mathbf{r}_2 \Rightarrow \mathbf{r}_2 \\ \\ 2\mathbf{r}_2 - \mathbf{r}_4 \Rightarrow \mathbf{r}_4 \end{array} \begin{bmatrix} -1 & 3 & 4 \\ 0 & 2 & 7 \\ 0 & -1 & 2 \\ 0 & 1 & 5 \end{bmatrix}$$

$$\Rightarrow \begin{array}{c} \\ \mathbf{r}_2 + 2\mathbf{r}_3 \Rightarrow \mathbf{r}_3 \\ \mathbf{r}_3 + \mathbf{r}_4 \Rightarrow \mathbf{r}_4 \end{array} \begin{bmatrix} -1 & 3 & 4 \\ 0 & 2 & 7 \\ 0 & 0 & 11 \\ 0 & 0 & 7 \end{bmatrix} \Rightarrow \begin{array}{c} \\ \\ \mathbf{r}_3 - \frac{11}{7}\mathbf{r}_4 \Rightarrow \mathbf{r}_4 \end{array} \begin{bmatrix} -1 & 3 & 4 \\ 0 & 2 & 7 \\ 0 & 0 & 11 \\ 0 & 0 & 0 \end{bmatrix}.$$

There are three nonzero pivots. Therefore, $\text{rank}(\mathbf{G}) = 3 = N$, and $\mathbf{G}$ is a full rank matrix.

∎

6. For a linear system of equations $\mathbf{Gx} = \mathbf{d}$, derive the least-squares solutions for $\mathbf{x}$, when $\mathbf{G}^{\mathrm{T}}\mathbf{G}$ is either non-singular or singular.

A: An objective function defined in the least-squares sense can be expressed as

$$\phi(\mathbf{x}) = \| \mathbf{d} - \mathbf{Gx} \|^2 .$$

Minimisation by setting

$$\frac{\partial \phi(\mathbf{x})}{\partial \mathbf{x}} = -2\mathbf{G}^T(\mathbf{d} - \mathbf{Gx}) = 0 ,$$

leads to the linear equation:

$$\mathbf{G}^T\mathbf{Gx} = \mathbf{G}^T\mathbf{d} .$$

If the square matrix $\mathbf{G}^T\mathbf{G}$ is non-singular, its inverse $[\mathbf{G}^T\mathbf{G}]^{-1}$ exists. The least-squares solution is

$$\mathbf{x} = [\mathbf{G}^T\mathbf{G}]^{-1}\mathbf{G}^T\mathbf{d} .$$

If $\mathbf{G}^T\mathbf{G}$ is a singular matrix, the inverse problem needs to be regularised. The objective function may be defined as

$$\phi(\mathbf{x}) = \| \mathbf{d} - \mathbf{Gx} \|^2 + \mu \| \mathbf{x} \|^2 ,$$

where μ is the regularisation factor. Hence, the least-squares solution is

$$\mathbf{x} = [\mathbf{G}^T\mathbf{G} + \mu\mathbf{I}]^{-1}\mathbf{G}^T\mathbf{d} .$$

Here, a small positive value μ added to the main diagonal of the square matrix $\mathbf{G}^T\mathbf{G}$ makes matrix $[\mathbf{G}^T\mathbf{G} + \mu\mathbf{I}]$ be invertible. ∎

7. Show that the least-squares problem for $\mathbf{Q}^T\mathbf{Gx} = \mathbf{Q}^T\mathbf{d}$, where $\mathbf{Q}$ is an orthogonal matrix, is the same as the least-squares problem for $\mathbf{Gx} = \mathbf{d}$.

A: The least-squares problem for $\mathbf{Q}^T\mathbf{Gx} = \mathbf{Q}^T\mathbf{d}$ is to minimise the (squared) L_2 norm $\| \mathbf{Q}^T\mathbf{d} - \mathbf{Q}^T\mathbf{Gx} \|^2$. The least-squares problem for $\mathbf{Gx} = \mathbf{d}$ is to minimise the (squared) L_2 norm $\| \mathbf{d} - \mathbf{Gx} \|^2$. Since $\mathbf{Q}$ is an orthogonal matrix, $\mathbf{QQ}^T = \mathbf{I}$, we can verify that

$$\| \mathbf{Q}^T\mathbf{d} - \mathbf{Q}^T\mathbf{Gx} \|^2 = (\mathbf{Q}^T\mathbf{d} - \mathbf{Q}^T\mathbf{Gx})^T(\mathbf{Q}^T\mathbf{d} - \mathbf{Q}^T\mathbf{Gx})$$

$$= (\mathbf{d} - \mathbf{Gx})^T\mathbf{QQ}^T(\mathbf{d} - \mathbf{Gx}) = (\mathbf{d} - \mathbf{Gx})^T(\mathbf{d} - \mathbf{Gx})$$

$$= \| \mathbf{d} - \mathbf{Gx} \|^2 .$$

Thus, for any orthogonal matrix $\mathbf{Q}$, this norm is unchanged by transformation of $\mathbf{Q}^T$. ∎

8. For a 2×2 matrix

$$\mathbf{A} = \begin{bmatrix} a & b \\ c & d \end{bmatrix},$$

calculate the eigenvalues, and prove that the product of eigenvalues is equal to the determinant, $\det |\mathbf{A}| = ad - bc$, and that the sum of the eigenvalues is equal to the trace, $\text{trace}(\mathbf{A}) = a + d$.

A: The characteristic equation is

$$\begin{vmatrix} a - \lambda & b \\ c & d - \lambda \end{vmatrix} = 0 .$$

Solving this equation

$$\lambda^2 - (a + d)\lambda + ad - bc = 0 .$$

Hence, the two eigenvalues are

$$\lambda_{1,2} = \frac{1}{2}\left(a + d \pm \sqrt{(a - d)^2 + 4bc} \right) .$$

Once we obtain these two eigenvalues, we prove that

$$\lambda_1 \lambda_2 = \frac{1}{4}\left\{ (a + d)^2 - [(a - d)^2 + 4bc] \right\} = ad - bc ,$$

and

$$\lambda_1 + \lambda_2 = a + d . \qquad \blacksquare$$

9.   For a 2×2 matrix

$$\mathbf{A} = \begin{bmatrix} 1 & 1 \\ 1 - \varepsilon & 1 \end{bmatrix} ,$$

calculate the eigenvalues of $\mathbf{A}$ and $\mathbf{A}^T\mathbf{A}$, and estimate the ratio of eigenvalues of $\mathbf{A}$ and the ratio of eigenvalues of $\mathbf{A}^T\mathbf{A}$, with $\varepsilon = 0.01$.

A: The eigenvalues of the square matrix $\mathbf{A}$ are

$$\lambda_{1,2} = 1 \pm \sqrt{1 - \varepsilon} .$$

The ratio of these two eigenvalues of $\mathbf{A}$ is

$$\frac{\lambda_1}{\lambda_2} = \frac{1 + \sqrt{1 - \varepsilon}}{1 - \sqrt{1 - \varepsilon}} \approx \frac{4}{\varepsilon} - 2 .$$

For $\varepsilon = 0.01$, it has an eigenvalue ratio $\lambda_1 / \lambda_2 \approx 398$.

For the eigenvalues of $\mathbf{A}^T\mathbf{A}$, we perform a transformation as follows:

$$\mathbf{A}^T\mathbf{A} = \begin{bmatrix} a_1 & b_1 \\ b_1 & a_2 \end{bmatrix} \quad \Rightarrow \quad \mathbf{P}(\mathbf{A}^T\mathbf{A})\mathbf{P}^T = \begin{bmatrix} \bar{a}_1 & 0 \\ 0 & \bar{a}_2 \end{bmatrix} ,$$

where **P** is an orthogonal matrix. If **P** is a clockwise rotation matrix, then **P**^{T} is a counterclockwise rotation matrix. We can write the transformation explicitly as

$$
\begin{bmatrix} c & s \\ -s & c \end{bmatrix} \begin{bmatrix} a_1 & b_1 \\ b_1 & a_2 \end{bmatrix} \begin{bmatrix} c & -s \\ s & c \end{bmatrix}
$$

$$
= \begin{bmatrix} a_1 c^2 + a_2 s^2 + 2b_1 cs & (a_2 - a_1)cs + b_1(c^2 - s^2) \\ (a_2 - a_1)cs + b_1(c^2 - s^2) & a_1 s^2 + a_2 c^2 - 2b_1 cs \end{bmatrix} = \begin{bmatrix} \bar{a}_1 & 0 \\ 0 & \bar{a}_2 \end{bmatrix}.
$$

The transformation makes the two off-diagonal elements to be zero. Therefore, we have a system of equations such as

$$
c^2 + s^2 = 1, \qquad (a_2 - a_1)cs + b_1(c^2 - s^2) = 0.
$$

The solution to this system is

$$
c^2 = \frac{1}{2}\left(1 - \frac{a_2 - a_1}{r}\right), \qquad s^2 = \frac{1}{2}\left(1 + \frac{a_2 - a_1}{r}\right), \qquad cs = \frac{b_1}{r},
$$

and

$$
r = \sqrt{(a_1 - a_2)^2 + 4b_1^2}.
$$

In the transformed matrix, the two diagonal elements are

$$
\bar{a}_1 = \frac{1}{2}(a_1 + a_2 + r), \qquad \bar{a}_2 = \frac{1}{2}(a_1 + a_2 - r).
$$

In this exercise,

$$
\mathbf{A}^T \mathbf{A} = \begin{bmatrix} 2 & 2-\varepsilon \\ 2-\varepsilon & 1+(1-\varepsilon)^2 \end{bmatrix},
$$

and

$$
\mathbf{P}(\mathbf{A}^T \mathbf{A})\mathbf{P}^T = \begin{bmatrix} 2-\varepsilon+\dfrac{\varepsilon^2}{2}+\dfrac{r}{2} & 0 \\ 0 & 2-\varepsilon+\dfrac{\varepsilon^2}{2}-\dfrac{r}{2} \end{bmatrix},
$$

where

$$
r = (2-\varepsilon)\sqrt{4+\varepsilon^2}.
$$

Therefore, the eigenvalues of matrix **A**^{T}**A** are

$$
\lambda_{1,2}^2 \equiv \bar{a}_{1,2} = 2-\varepsilon+\frac{\varepsilon^2}{2} \pm \frac{r}{2}.
$$

The ratio of these two eigenvalues is

$$
\frac{\lambda_1^2}{\lambda_2^2} \equiv \frac{\bar{a}_1}{\bar{a}_2} = \frac{4-2\varepsilon+\varepsilon^2+r}{4-2\varepsilon+\varepsilon^2-r}.
$$

For $\varepsilon = 0.01$, $r \approx 3.98004975$, $\bar{a}_1 \approx 3.98$, $\bar{a}_2 \approx 2.4875 \times 10^{-5}$, we have the ratio $\lambda_1^2 / \lambda_2^2 \approx 1.6 \times 10^5$.   ∎

10. For a 2×2 matrix, $\mathbf{A} = \begin{bmatrix} 1 & 2 \\ 3 & 4 \end{bmatrix}$, calculate the condition numbers of $\mathbf{A}$ and

$\mathbf{A}^T\mathbf{A}$, based on the eigenvalues.

A: The characteristic equation for $\mathbf{A}$ is

$$\begin{vmatrix} 1-\lambda & 2 \\ 3 & 4-\lambda \end{vmatrix} = 0.$$

Solving this equation, $\lambda^2 - 5\lambda - 2 = 0$, we obtain two eigenvalues

$$\lambda_{1,2} = \frac{5 \pm \sqrt{33}}{2}.$$

Hence, the condition number of $\mathbf{A}$ is $|\lambda_1 / \lambda_2| = \frac{1}{4}(29 + 5 \times \sqrt{33}) \approx 14.43$.

For $\mathbf{A}^T\mathbf{A} = \begin{bmatrix} 10 & 14 \\ 14 & 20 \end{bmatrix}$, the characteristic equation is

$$\begin{vmatrix} 10-\lambda^2 & 14 \\ 14 & 20-\lambda^2 \end{vmatrix} = 0.$$

Solving this equation $\lambda^4 - 30\lambda^2 + 4 = 0$, we obtain two eigenvalues

$$\lambda_{1,2}^2 = 15 \pm \sqrt{221}.$$

The condition number of $\mathbf{A}^T\mathbf{A}$ is $\lambda_1^2 / \lambda_2^2 = \frac{1}{2}(223 + 15 \times \sqrt{221}) \approx 223$.

Alternatively, one can use the following formulae, presented in the previous exercise:

$$r = \sqrt{(a_1 - a_2)^2 + 4b_1^2},$$

$$\bar{a}_{1,2} = \frac{1}{2}(a_1 + a_2 \pm r).$$

In this case, for the symmetric matrix with elements $a_1 = 10$, $a_2 = 20$, $b_1 = 14$, we have

$$r = 2 \times \sqrt{221},$$

$$\lambda_{1,2}^2 \equiv \bar{a}_{1,2} = 15 \pm \sqrt{221}.$$

The condition number of $\mathbf{A}^T\mathbf{A}$ is $\lambda_1^2 / \lambda_2^2 \approx 223$.   ∎

11. Given a 2×3 matrix

$$\mathbf{G} = \begin{bmatrix} 1 & 1 & 0 \\ 0 & 0 & 1 \end{bmatrix},$$

   (1) find the eigenvalues of $\mathbf{G}^\mathrm{T}\mathbf{G}$ and $\mathbf{G}\mathbf{G}^\mathrm{T}$,

   (2) perform SVD, and

   (3) verify $\mathbf{G} = \mathbf{U}\boldsymbol{\Lambda}\mathbf{V}^\mathrm{T}$.

A: (1) To find the eigenvalues of the square matrix

$$\mathbf{G}^\mathrm{T}\mathbf{G} = \begin{bmatrix} 1 & 1 & 0 \\ 1 & 1 & 0 \\ 0 & 0 & 1 \end{bmatrix},$$

we need to find the roots of the characteristic polynomial:

$$\det(\mathbf{G}^\mathrm{T}\mathbf{G}) = \begin{vmatrix} 1-\lambda^2 & 1 & 0 \\ 1 & 1-\lambda^2 & 0 \\ 0 & 0 & 1-\lambda^2 \end{vmatrix} = 0,$$

$$(1-\lambda^2)^3 - (1-\lambda^2) = \lambda^2(1-\lambda^2)(\lambda^2 - 2) = 0.$$

Thus, three eigenvalues are

$$\lambda_1^2 = 2, \quad \lambda_2^2 = 1, \quad \lambda_3^2 = 0.$$

Now, let us find the eigenvalues of $\mathbf{G}\mathbf{G}^\mathrm{T}$:

$$\mathbf{G}\mathbf{G}^\mathrm{T} = \begin{bmatrix} 2 & 0 \\ 0 & 1 \end{bmatrix}.$$

Since $\mathbf{G}\mathbf{G}^\mathrm{T}$ is a diagonal matrix, the diagonal entries are the eigenvalues:

$$\det(\mathbf{G}\mathbf{G}^\mathrm{T}) = \begin{vmatrix} 2-\lambda^2 & 0 \\ 0 & 1-\lambda^2 \end{vmatrix} = 0,$$

and

$$\lambda_1^2 = 2, \text{ and } \lambda_2^2 = 1.$$

(2) First, compute the eigenvectors $\mathbf{u}_i$ by solving the eigenvalue problem

$$\mathbf{G}\mathbf{G}^\mathrm{T}\mathbf{u}_i = \lambda_i^2\mathbf{u}_i,$$

for $\lambda_1^2 = 2$, $\lambda_2^2 = 1$. Thus,

$$\begin{bmatrix} 2 & 0 \\ 0 & 1 \end{bmatrix}\begin{bmatrix} u_{11} \\ u_{21} \end{bmatrix} = 2\begin{bmatrix} u_{11} \\ u_{21} \end{bmatrix}, \quad \mathbf{u}_1 = \begin{bmatrix} 1 \\ 0 \end{bmatrix},$$

$$\begin{bmatrix} 2 & 0 \\ 0 & 1 \end{bmatrix} \begin{bmatrix} u_{11} \\ u_{21} \end{bmatrix} = \begin{bmatrix} u_{11} \\ u_{21} \end{bmatrix}, \qquad \mathbf{u}_2 = \begin{bmatrix} 0 \\ 1 \end{bmatrix},$$

and

$$\mathbf{U} = \begin{bmatrix} 1 & 0 \\ 0 & 1 \end{bmatrix}.$$

Then, compute the eigenvectors $\mathbf{v}_i$ by solving the eigenvalue problem

$$\mathbf{G}^T\mathbf{G}\mathbf{v}_i = \lambda_i^2\mathbf{v}_i, \quad \text{for } \lambda_1^2 = 2, \ \lambda_2^2 = 1, \ \lambda_3^2 = 0.$$

For $\lambda_1^2 = 2$,

$$\begin{bmatrix} 1 & 1 & 0 \\ 1 & 1 & 0 \\ 0 & 0 & 1 \end{bmatrix} \begin{bmatrix} v_{11} \\ v_{21} \\ v_{31} \end{bmatrix} = 2 \begin{bmatrix} v_{11} \\ v_{21} \\ v_{31} \end{bmatrix},$$

which means that $v_{11} = v_{21}$, and $v_{31} = 0$. The normalised singular vector is

$$\mathbf{v}_1 = \begin{bmatrix} 1/\sqrt{2} \\ 1/\sqrt{2} \\ 0 \end{bmatrix}.$$

For $\lambda_2^2 = 1$,

$$\begin{bmatrix} 1 & 1 & 0 \\ 1 & 1 & 0 \\ 0 & 0 & 1 \end{bmatrix} \begin{bmatrix} v_{12} \\ v_{22} \\ v_{32} \end{bmatrix} = \begin{bmatrix} v_{12} \\ v_{22} \\ v_{32} \end{bmatrix}, \quad \mathbf{v}_2 = \begin{bmatrix} 0 \\ 0 \\ 1 \end{bmatrix}.$$

For $\lambda_3^2 = 0$, $\mathbf{G}^T\mathbf{G}\mathbf{v}_3 = \mathbf{0}$, then we must solve $\mathbf{G}\mathbf{v}_3 = \mathbf{0}$ to find the null space singular vector.

$$\begin{bmatrix} 1 & 1 & 0 \\ 0 & 0 & 1 \end{bmatrix} \begin{bmatrix} v_{13} \\ v_{23} \\ v_{33} \end{bmatrix} = \begin{bmatrix} 0 \\ 0 \end{bmatrix}.$$

This means that $v_{13} + v_{23} = 0$ and $v_{33} = 0$, hence the normalised model null space singular vector is

$$\mathbf{v}_3 = \begin{bmatrix} -1/\sqrt{2} \\ 1/\sqrt{2} \\ 0 \end{bmatrix}.$$

Thus, these give us the model space singular vectors in a matrix form as

$$\mathbf{V} = \begin{bmatrix} 1/\sqrt{2} & 0 & -1/\sqrt{2} \\ 1/\sqrt{2} & 0 & 1/\sqrt{2} \\ 0 & 1 & 1 \end{bmatrix}.$$

(3) The singular value matrix is

$$\mathbf{\Lambda} = \begin{bmatrix} \sqrt{2} & 0 & 0 \\ 0 & 1 & 0 \end{bmatrix}.$$

We can verify the SVD directly:

$$\mathbf{G} = \mathbf{U}\mathbf{\Lambda}\mathbf{V}^{\mathrm{T}} = \begin{bmatrix} 1 & 0 \\ 0 & 1 \end{bmatrix}\begin{bmatrix} \sqrt{2} & 0 & 0 \\ 0 & 1 & 0 \end{bmatrix}\begin{bmatrix} 1/\sqrt{2} & 1/\sqrt{2} & 0 \\ 0 & 0 & 1 \\ -1/\sqrt{2} & 1/\sqrt{2} & 0 \end{bmatrix}$$

$$= \begin{bmatrix} 1 & 1 & 0 \\ 0 & 0 & 1 \end{bmatrix}.$$ ∎

12. For $\mathbf{A} = \begin{bmatrix} 2 & 1 \\ 1 & 2 \end{bmatrix}$, compute $\|\mathbf{A}\|_1$, $\|\mathbf{A}\|_2$, and $\|\mathbf{A}\|_\infty$.

A: The L_1 norm is the maximum value among the sums of the absolute values of the columns:

$$\|\mathbf{A}\|_1 = \max_{1\le j\le N}\sum_{i=1}^{N}|a_{ij}| = 3.$$

For L_2 norm, we need to calculate the largest eigenvalue of the matrix. The characteristic equation for $\mathbf{A}$ is

$$\begin{vmatrix} 2-\lambda & 1 \\ 1 & 2-\lambda \end{vmatrix} = 0.$$

Solving this equation by factorisation:

$$(\lambda-3)(\lambda-1) = 0,$$

$$\lambda_1 = 3, \quad \text{and} \quad \lambda_2 = 1,$$

so that, $\|\mathbf{A}\|_2 = 3$.

The L_∞ norm (the infinity-norm) is the maximum value among the sums of the absolute values of the rows:

$$\|\mathbf{A}\|_\infty = \max_{1\le i\le N}\sum_{j=1}^{N}|a_{ij}| = 3.$$

In this special case, we have $\| A \|_1 = \| A \|_2 = \| A \|_\infty = 3$, which satisfies the property:

$$\| A \|_2^2 \leq \| A \|_1 \| A \|_\infty .$$

∎

13. In the steepest descent method, the step length α is defined by $\alpha = (\mathbf{e}, \mathbf{e})/(\mathbf{e}, \mathbf{A}\mathbf{e})$, where $\mathbf{e} = \mathbf{b} - \mathbf{A}\mathbf{x}$ is the residual vector. Verify

$$\phi(\mathbf{x} + 2\alpha \mathbf{e}) = \phi(\mathbf{x}) .$$

A: As the error measurement $\phi(\mathbf{x})$ is a quadratic function in variable vector $\mathbf{x}$:

$$\phi(\mathbf{x}) = (\mathbf{x}, \mathbf{A}\mathbf{x}) - 2(\mathbf{b}, \mathbf{x}) + (\mathbf{b}, \mathbf{A}^{-1}\mathbf{b}) ,$$

we have

$$\phi(\mathbf{x}^{(k)} + 2\alpha \mathbf{e}^{(k)})$$

$$= ((\mathbf{x}^{(k)} + 2\alpha \mathbf{e}^{(k)}), \mathbf{A}(\mathbf{x}^{(k)} + 2\alpha \mathbf{e}^{(k)})) \ - 2(\mathbf{b}, (\mathbf{x}^{(k)} + 2\alpha \mathbf{e}^{(k)})) + (\mathbf{b}, \mathbf{A}^{-1}\mathbf{b})$$

$$= \phi(\mathbf{x}^{(k)}) + 4\alpha^2(\mathbf{e}^{(k)}, \mathbf{A}\mathbf{e}^{(k)}) + 2\alpha(\mathbf{x}^{(k)}, \mathbf{A}\mathbf{e}^{(k)}) \ + 2\alpha(\mathbf{e}^{(k)}, \mathbf{A}\mathbf{x}^{(k)}) - 4\alpha(\mathbf{b}, \mathbf{e}^{(k)})$$

$$= \phi(\mathbf{x}^{(k)}) + 4\alpha^2(\mathbf{e}^{(k)}, \mathbf{A}\mathbf{e}^{(k)}) + 4\alpha(\mathbf{e}^{(k)}, \mathbf{A}\mathbf{x}^{(k)}) - 4\alpha(\mathbf{b}, \mathbf{e}^{(k)})$$

$$= \phi(\mathbf{x}^{(k)}) + 4\alpha^2(\mathbf{e}^{(k)}, \mathbf{A}\mathbf{e}^{(k)}) - 4\alpha(\mathbf{e}^{(k)}, \mathbf{e}^{(k)}) .$$

Because $\alpha = (\mathbf{e}^{(k)}, \mathbf{e}^{(k)})/(\mathbf{e}^{(k)}, \mathbf{A}\mathbf{e}^{(k)})$, we obtain

$$\phi(\mathbf{x}^{(k)} + 2\alpha \mathbf{e}^{(k)}) = \phi(\mathbf{x}^{(k)}) .$$

Therefore, $\mathbf{x}^{(k)}$ and $\mathbf{x}^{(k)} + 2\alpha \mathbf{e}^{(k)}$ are on the same contour $\phi(\mathbf{x}^{(k)})$, and $\mathbf{x}^{(k+1)} = \mathbf{x}^{(k)} + \alpha \mathbf{e}^{(k)}$ is the midpoint in between.   ∎

14. An error function is defined by the inner product $\phi(\mathbf{e}) = (\mathbf{e}, \mathbf{A}^{-1}\mathbf{e})$, where $\mathbf{e}$ is the residual vector, and $\mathbf{A}$ is a positive definite symmetric matrix. Using the definition of step length $\alpha = (\mathbf{e}^{(k)}, \mathbf{e}^{(k)})/(\mathbf{e}^{(k)}, \mathbf{A}\mathbf{e}^{(k)})$, where k is the iteration number, prove that the steepest descent method is converged.

A: Substituting $\mathbf{e} = \mathbf{b} - \mathbf{A}\mathbf{x}$ into the error function leads to

$$\phi(\mathbf{x}) = (\mathbf{x}, \mathbf{A}\mathbf{x}) - 2(\mathbf{b}, \mathbf{x}) + (\mathbf{b}, \mathbf{A}^{-1}\mathbf{b}) .$$

The difference of the error functions between two consecutive iterations is

$$\phi(\mathbf{x}^{(k+1)}) - \phi(\mathbf{x}^{(k)}) = (\mathbf{x}^{(k+1)}, \mathbf{A}\mathbf{x}^{(k+1)}) - (\mathbf{x}^{(k)}, \mathbf{A}\mathbf{x}^{(k)}) - 2(\mathbf{b}, \mathbf{x}^{(k+1)} - \mathbf{x}^{(k)}) . \quad \text{(E.1)}$$

Substituting $\mathbf{x}^{(k+1)} = \mathbf{x}^{(k)} + \alpha \mathbf{e}^{(k)}$, the first inner product is

$$(\mathbf{x}^{(k+1)}, \mathbf{A}\mathbf{x}^{(k+1)}) = (\mathbf{x}^{(k)}, \mathbf{A}\mathbf{x}^{(k)}) + 2\alpha(\mathbf{e}^{(k)}, \mathbf{A}\mathbf{x}^{(k)}) + \alpha^2(\mathbf{e}^{(k)}, \mathbf{A}\mathbf{e}^{(k)}) .$$

Here, matrix $\mathbf{A}$ is assumed to be symmetric. The last inner product in Equation E.1 is

$$(\mathbf{b}, \mathbf{x}^{(k+1)} - \mathbf{x}^{(k)}) = (\mathbf{b}, \alpha \mathbf{e}^{(k)}) = \alpha(\mathbf{A}\mathbf{x}^{(k)}, \mathbf{e}^{(k)}) + \alpha(\mathbf{e}^{(k)}, \mathbf{e}^{(k)}) \,.$$

Then, the difference of Equation E.1 becomes

$$\phi(\mathbf{x}^{(k+1)}) - \phi(\mathbf{x}^{(k)}) = \alpha^2(\mathbf{e}^{(k)}, \mathbf{A}\mathbf{e}^{(k)}) - 2\alpha(\mathbf{e}^{(k)}, \mathbf{e}^{(k)}) \,.$$

Using definition of $\alpha = (\mathbf{e}^{(k)}, \mathbf{e}^{(k)}) / (\mathbf{e}^{(k)}, \mathbf{A}\mathbf{e}^{(k)})$ leads to

$$\phi(\mathbf{x}^{(k+1)}) - \phi(\mathbf{x}^{(k)}) = -\frac{(\mathbf{e}^{(k)}, \mathbf{e}^{(k)})^2}{(\mathbf{e}^{(k)}, \mathbf{A}\mathbf{e}^{(k)})} \le 0 \,,$$

where $(\mathbf{e}^{(k)}, \mathbf{A}\mathbf{e}^{(k)}) > 0$, for positive definite matrix $\mathbf{A}$ and for all nonzero real vectors $\mathbf{e}$.

Since $\phi(\mathbf{x}^{(k+1)}) \le \phi(\mathbf{x}^{(k)})$, for any $\mathbf{x}^{(k)}$, the steepest descent method must converge. The proof is complete. ∎

15. For solving the inverse problem $\mathbf{A}\mathbf{x} = \mathbf{b}$, an iterative method defines the error function in a quadratic form, as

$$\phi(\mathbf{x}) = (\mathbf{x}, \mathbf{A}\mathbf{x}) - 2(\mathbf{b}, \mathbf{x}) + (\mathbf{b}, \mathbf{A}^{-1}\mathbf{b}) \,.$$

The steepest descent method repeatedly minimises $\phi(\mathbf{x})$ along directions defined by the residual vector $\mathbf{e} = \mathbf{b} - \mathbf{A}\mathbf{x}$. Prove that this procedure is a gradient-based method, and when the gradient $\phi'(\mathbf{x})$ vanishes, that $\mathbf{x}$ is the solution of $\mathbf{A}\mathbf{x} = \mathbf{b}$.

A: The gradient of the error function $\phi(\mathbf{x})$ is

$$\phi'(\mathbf{x}) \equiv \frac{\partial \phi}{\partial \mathbf{x}} = 2(\mathbf{A}\mathbf{x} - \mathbf{b}) \,.$$

The residual vector $\mathbf{e}$ is proportional to the maximum gradient:

$$\mathbf{e} \equiv \mathbf{b} - \mathbf{A}\mathbf{x} = -\frac{1}{2}\phi'(\mathbf{x}) \,.$$

This proves that the steepest descent method updates the solution estimate along the negative gradient direction.

If $\phi'(\mathbf{x}) = \mathbf{0}$, then $\mathbf{e} \equiv \mathbf{b} - \mathbf{A}\mathbf{x} = \mathbf{0}$. Therefore, $\mathbf{x}$ is the solution of equation $\mathbf{A}\mathbf{x} = \mathbf{b}$. ∎

16. Assume that a solution $\mathbf{x}$ to $\mathbf{G}\mathbf{x} = \mathbf{d}$ exists, and prove

$$\| \mathbf{x} \|^2 = \sum_k \lambda_k^{-2}(\mathbf{d}, \mathbf{u}_k)^2 \,,$$

where $\{\lambda_k^2\}$ are the eigenvalues of $\mathbf{G}\mathbf{G}^{\mathrm{T}}$, $\{\mathbf{u}_k\}$ are the associated eigenvectors, and $(\mathbf{d}, \mathbf{u}_k)$ is the inner product of two vectors.

A: First, based on singular value decomposition theory, the eigenvectors $\{\mathbf{v}_k\}$, associated with $\mathbf{G}^T\mathbf{G}$, form an orthonormal basis for 'model vector' $\mathbf{x}$. Then, we can write $\mathbf{x}$ as the sum of the convergent series of $(\mathbf{x}, \mathbf{v}_k)\mathbf{v}_k$ and $\parallel \mathbf{x} \parallel^2$ as

$$\parallel \mathbf{x} \parallel^2 = \sum_k (\mathbf{x}, \mathbf{v}_k)^2 . \tag{E.2}$$

Secondly, from the definitions of eigenvectors, $\mathbf{G}\mathbf{v} = \lambda\mathbf{u}$ and $\mathbf{G}^T\mathbf{u} = \lambda\mathbf{v}$, then

$$\mathbf{v} = \lambda^{-2}\mathbf{G}^T\mathbf{G}\mathbf{v} ,$$

we may express (E.2) as

$$\parallel \mathbf{x} \parallel^2 = \sum_k \lambda_k^{-4}(\mathbf{x}, \mathbf{G}^T\mathbf{G}\mathbf{v}_k)^2 . \tag{E.3}$$

Thirdly, assume that a solution $\mathbf{x}$ to $\mathbf{G}\mathbf{x} = \mathbf{d}$ exists, then $\mathbf{G}^T\mathbf{G}\mathbf{x} = \mathbf{G}^T\mathbf{d}$. Let us calculate the scalar product of both parts of this equality and $\mathbf{v}_k$,

$$(\mathbf{G}^T\mathbf{G}\mathbf{x}, \mathbf{v}_k) = (\mathbf{G}^T\mathbf{d}, \mathbf{v}_k) = (\mathbf{d}, \mathbf{G}\mathbf{v}_k) .$$

From the definition of $\mathbf{u}_k$, $\mathbf{G}\mathbf{v} = \lambda\mathbf{u}$, we obtain

$$(\mathbf{G}^T\mathbf{G}\mathbf{x}, \mathbf{v}_k) = \lambda_k(\mathbf{d}, \mathbf{u}_k) . \tag{E.4}$$

Finally, substituting Equation E.4 into Equation E.3, we prove

$$\parallel \mathbf{x} \parallel^2 = \sum_k \lambda_k^{-2}(\mathbf{d}, \mathbf{u}_k)^2 .$$

Alternatively, we can use the matrix form of SVD, $\mathbf{G} = \mathbf{U}\mathbf{\Lambda}\mathbf{V}^T$. Assume that a solution $\mathbf{x}$ to $\mathbf{G}\mathbf{x} = \mathbf{d}$ exists, then

$$\mathbf{x} = \mathbf{V}\mathbf{\Lambda}^{-1}\mathbf{U}^T\mathbf{d} .$$

Now, we have

$$\parallel \mathbf{x} \parallel^2 = [\mathbf{V}\mathbf{\Lambda}^{-1}\mathbf{U}^T\mathbf{d}]^T \mathbf{V}\mathbf{\Lambda}^{-1}\mathbf{U}^T\mathbf{d}$$

$$= \mathbf{d}^T\mathbf{U}\mathbf{\Lambda}^{-2}\mathbf{U}^T\mathbf{d}$$

$$= \mathbf{\Lambda}^{-2}[\mathbf{U}^T\mathbf{d}]^2 = \sum_k \lambda_k^{-2}(\mathbf{d}, \mathbf{u}_k)^2.$$

The proof is complete.   ■

17. In the conjugate gradient method, the step length α is defined as

$$\alpha = \frac{(\mathbf{p}^{(k)}, \mathbf{e}^{(k)})}{(\mathbf{p}^{(k)}, \mathbf{A}\mathbf{p}^{(k)})}.$$

Prove that it is equal to the following definition as

$$\alpha = \frac{(\mathbf{e}^{(k)}, \mathbf{e}^{(k)})}{(\mathbf{p}^{(k)}, \mathbf{A}\mathbf{p}^{(k)})} .$$

A: To prove these two definitions of the step length α are equivalent, we just need to prove $(\mathbf{p}^{(k)}, \mathbf{e}^{(k)}) = (\mathbf{e}^{(k)}, \mathbf{e}^{(k)})$.

Assuming $\mathbf{z}$ is the true solution of the linear system $\mathbf{Ax} = \mathbf{b}$. For the given $\mathbf{x}^{(1)}$, the initial approximation to the solution, difference $\mathbf{z} - \mathbf{x}^{(1)}$ can be expanded in terms of the linearly independent search vectors as

$$\mathbf{z} - \mathbf{x}^{(1)} = \sum_{j=1}^{N} \xi_j \mathbf{p}^{(j)}.$$

Pre-multiplying both sides with matrix $\mathbf{A}$, we have

$$\mathbf{e}^{(1)} \equiv \mathbf{Az} - \mathbf{Ax}^{(1)} = \sum_{j=1}^{N} \xi_j \mathbf{Ap}^{(j)},$$

because $\mathbf{b} = \mathbf{Az}$. The process of building up $\mathbf{x}$ component by component can also be viewed as a process of cutting down the residual term component by component. After N iterations, every component is cut away, and the residual $\mathbf{e}^{(N)} = 0$. For the kth iteration, the residual is

$$\mathbf{e}^{(k)} = \sum_{j=k}^{N} \xi_j \mathbf{Ap}^{(j)}.$$

Taking the inner product with $\mathbf{p}^{(i)}$, for $i < k$, and considering $(\mathbf{p}^{(i)}, \mathbf{Ap}^{(j)}) = 0$, if $i \neq j$, we get

$$(\mathbf{p}^{(i)}, \mathbf{e}^{(k)}) = \sum_{j=k}^{N} \xi_j (\mathbf{p}^{(i)}, \mathbf{Ap}^{(j)}) = 0, \quad \text{for } i < k. \tag{E.5}$$

Now, writing down the equation for updating the search direction:

$$\mathbf{p}^{(k)} = \mathbf{e}^{(k)} + \beta \mathbf{p}^{(k-1)},$$

and taking the inner product with $\mathbf{e}^{(k)}$, we have

$$(\mathbf{p}^{(k)}, \mathbf{e}^{(k)}) = (\mathbf{e}^{(k)}, \mathbf{e}^{(k)}) + \beta (\mathbf{p}^{(k-1)}, \mathbf{e}^{(k)}).$$

Because of Equation E.5, $(\mathbf{p}^{(k-1)}, \mathbf{e}^{(k)}) = 0$, we obtain equation

$$(\mathbf{p}^{(k)}, \mathbf{e}^{(k)}) = (\mathbf{e}^{(k)}, \mathbf{e}^{(k)}).$$

Therefore, we prove in turn the step length α as

$$\alpha \equiv \frac{(\mathbf{p}^{(k)}, \mathbf{e}^{(k)})}{(\mathbf{p}^{(k)}, \mathbf{Ap}^{(k)})} = \frac{(\mathbf{e}^{(k)}, \mathbf{e}^{(k)})}{(\mathbf{p}^{(k)}, \mathbf{Ap}^{(k)})}.$$

The proof is complete.   ∎

18. In the conjugate gradient method, the conjugate coefficient is

$$\beta = -\frac{(\mathbf{e}^{(k+1)}, \mathbf{Ap}^{(k)})}{(\mathbf{p}^{(k)}, \mathbf{Ap}^{(k)})}.$$

Prove that it is equal to the following definition:

$$\beta = \frac{(\mathbf{e}^{(k+1)}, \mathbf{e}^{(k+1)})}{(\mathbf{e}^{(k)}, \mathbf{e}^{(k)})}.$$

A: In the conjugate gradient method, the search direction is updated by

$$\mathbf{p}^{(k)} = \mathbf{e}^{(k)} + \beta \mathbf{p}^{(k-1)},$$

Taking the inner product with $\mathbf{Ap}^{(k)}$,

$$(\mathbf{p}^{(k)}, \mathbf{Ap}^{(k)}) = (\mathbf{e}^{(k)}, \mathbf{Ap}^{(k)}) + \beta(\mathbf{p}^{(k-1)}, \mathbf{Ap}^{(k)}),$$

and because of the conjugacy constraint $(\mathbf{p}^{(k-1)}, \mathbf{Ap}^{(k)}) = 0$, we have the property,

$$(\mathbf{p}^{(k)}, \mathbf{Ap}^{(k)}) = (\mathbf{e}^{(k)}, \mathbf{Ap}^{(k)}). \tag{E.6}$$

The residual is updated by

$$\mathbf{e}^{(k+1)} = \mathbf{e}^{(k)} - \alpha \mathbf{Ap}^{(k)}. \tag{E.7}$$

Take the inner product with $\mathbf{e}^{(k)}$, and use Equation E.6,

$$(\mathbf{e}^{(k)}, \mathbf{e}^{(k+1)}) = (\mathbf{e}^{(k)}, \mathbf{e}^{(k)}) - \alpha_k(\mathbf{e}^{(k)}, \mathbf{Ap}^{(k+1)})$$

$$= (\mathbf{e}^{(k)}, \mathbf{e}^{(k)}) - \alpha_k(\mathbf{p}^{(k)}, \mathbf{Ap}^{(k+1)}).$$

Substituting $\alpha = (\mathbf{e}^{(k)}, \mathbf{e}^{(k)})/(\mathbf{p}^{(k)}, \mathbf{Ap}^{(k)})$, we have

$$(\mathbf{e}^{(k)}, \mathbf{e}^{(k+1)}) = 0. \tag{E.8}$$

Now, rewriting Equation E.7 as

$$\mathbf{Ap}^{(k)} = \frac{1}{\alpha}(\mathbf{e}^{(k)} - \mathbf{e}^{(k+1)}).$$

and taking the inner product with $\mathbf{e}^{(k+1)}$, we have

$$(\mathbf{e}^{(k+1)}, \mathbf{Ap}^{(k)}) = \frac{1}{\alpha}[(\mathbf{e}^{(k+1)}, \mathbf{e}^{(k)}) - (\mathbf{e}^{(k+1)}, \mathbf{e}^{(k+1)})] = -\frac{(\mathbf{e}^{(k+1)}, \mathbf{e}^{(k+1)})}{\alpha}$$

$$= -\frac{(\mathbf{e}^{(k+1)}, \mathbf{e}^{(k+1)})}{(\mathbf{e}^{(k)}, \mathbf{e}^{(k)})}(\mathbf{p}^{(k)}, \mathbf{Ap}^{(k)}). \tag{E.9}$$

because $(\mathbf{e}^{(k+1)}, \mathbf{e}^{(k)}) = 0$ (Equation E.8) and $\alpha = (\mathbf{e}^{(k)}, \mathbf{e}^{(k)})/(\mathbf{p}^{(k)}, \mathbf{Ap}^{(k)})$. By rewriting Equation E.9, we prove that

$$\beta \equiv -\frac{(\mathbf{e}^{(k+1)}, \mathbf{A}\mathbf{p}^{(k)})}{(\mathbf{p}^{(k)}, \mathbf{A}\mathbf{p}^{(k)})} = \frac{(\mathbf{e}^{(k+1)}, \mathbf{e}^{(k+1)})}{(\mathbf{e}^{(k)}, \mathbf{e}^{(k)})}.$$

The proof is complete. ∎

19. In the complex biconjugate gradient method, the biconjugate coefficient is defined by

$$\beta = -\frac{(\mathbf{A}^{H}\hat{\mathbf{p}}^{(k)}, \mathbf{e}^{(k+1)})}{(\hat{\mathbf{p}}^{(k)}, \mathbf{A}\mathbf{p}^{(k)})},$$

where $\mathbf{A}^{H}$ denotes the Hermitian transpose of the complex matrix $\mathbf{A}$. Prove this biconjugate coefficient can also be expressed as

$$\beta = \frac{(\hat{\mathbf{e}}^{(k+1)}, \mathbf{e}^{(k+1)})}{(\hat{\mathbf{e}}^{(k)}, \mathbf{e}^{(k)})}.$$

where the residual and the biresidual are given by

$$\mathbf{e}^{(k+1)} = \mathbf{e}^{(k)} - \alpha \mathbf{A}\mathbf{p}^{(k)},$$

$$\hat{\mathbf{e}}^{(k+1)} = \hat{\mathbf{e}}^{(k)} - \bar{\alpha}\mathbf{A}^{H}\hat{\mathbf{p}}^{(k)},$$

and $\bar{\alpha}$ is the complex conjugate of α.

A: Taking the inner product between $\hat{\mathbf{e}}^{(k+1)}$ and $\mathbf{e}^{(i)}$ as the following,

$$(\hat{\mathbf{e}}^{(k+1)}, \mathbf{e}^{(i)}) = (\hat{\mathbf{e}}^{(k)} - \bar{\alpha}\mathbf{A}^{H}\hat{\mathbf{p}}^{(k)}, \mathbf{e}^{(i)}) = (\hat{\mathbf{e}}^{(k)}, \mathbf{e}^{(i)}) - (\bar{\alpha}\mathbf{A}^{H}\hat{\mathbf{p}}^{(k)}, \mathbf{e}^{(i)})$$

$$= (\hat{\mathbf{e}}^{(k)}, \mathbf{e}^{(i)}) - \alpha(\mathbf{A}^{H}\hat{\mathbf{p}}^{(k)}, \mathbf{e}^{(i)}).$$

Note here, that the inner product $(\bar{\alpha}\mathbf{x}, \mathbf{y}) = \alpha \mathbf{x}^{H}\mathbf{y} = \alpha(\mathbf{x}, \mathbf{y})$, where α is a complex scalar. We have

$$\alpha(\mathbf{A}^{H}\hat{\mathbf{p}}^{(k)}, \mathbf{e}^{(i)}) = (\hat{\mathbf{e}}^{(k)}, \mathbf{e}^{(i)}) - (\hat{\mathbf{e}}^{(k+1)}, \mathbf{e}^{(i)}).$$

That is,

$$\alpha(\mathbf{A}^{H}\hat{\mathbf{p}}^{(k)}, \mathbf{e}^{(i)}) = \begin{cases} (\hat{\mathbf{e}}^{(k)}, \mathbf{e}^{(k)}), & i = k, \\ -(\hat{\mathbf{e}}^{(k+1)}, \mathbf{e}^{(k+1)}), & i = k+1, \\ 0, & \text{otherwise.} \end{cases}$$

Using the equation for $i = k+1$, we have

$$(\mathbf{A}^{H}\hat{\mathbf{p}}^{(k)}, \mathbf{e}^{(k+1)}) = -\frac{1}{\alpha}(\hat{\mathbf{e}}^{(k+1)}, \mathbf{e}^{(k+1)}). \tag{E.10}$$

In the complex biconjugate gradient method, the step-length parameter is defined by $\alpha = (\hat{\mathbf{e}}^{(k)}, \mathbf{e}^{(k)})/(\hat{\mathbf{p}}^{(k)}, \mathbf{A}\mathbf{p}^{(k)})$. That is,

$$(\hat{\mathbf{p}}^{(k)}, \mathbf{A}\mathbf{p}^{(k)}) = \frac{1}{\alpha}(\hat{\mathbf{e}}^{(k)}, \mathbf{e}^{(k)}). \tag{E.11}$$

Therefore, using Equations E.10 and E.11, we prove that

$$\beta \equiv -\frac{(\mathbf{A}^H\hat{\mathbf{p}}^{(k)}, \mathbf{e}^{(k+1)})}{(\hat{\mathbf{p}}^{(k)}, \mathbf{A}\mathbf{p}^{(k)})} = \frac{(\hat{\mathbf{e}}^{(k+1)}, \mathbf{e}^{(k+1)})}{(\hat{\mathbf{e}}^{(k)}, \mathbf{e}^{(k)})}.$$

The proof is complete.   ∎

20. In seismic inversion, the objective function is often defined as

$$\phi = Q + \mu R,$$

where Q is the data fitting quality, R is a model regularisation, and μ is the trade-off parameter that balances the data matching and the model regularisation.

(1) What is Tikhonov regularisation? Analyse the difference between a standard L_2 norm model constraint and a general Tikhonov regularisation.

(2) What is the maximum entropy regularisation? Analyse the difference between the standard L_2 norm model constraint and the maximum entropy regularisation.

(3) What is Cauchy regularisation? What is the main purpose of Cauchy distribution regularisation, as used in seismic inversion?

A: (1) The Tikhonov regularisation is defined as

$$R = \int_a^b \left(p \left\| \frac{\partial \mathbf{x}}{\partial y} \right\|^2 + q \left\| \mathbf{x} - \mathbf{x}_{\text{ref}} \right\|^2 \right) dy,$$

where p and q are constant. A standard L_2 norm model regularisation is

$$R = \left\| \mathbf{x} - \mathbf{x}_{\text{ref}} \right\|^2,$$

where $\mathbf{x}_{\text{ref}}$ is a reference model. The Tikhonov regularisation has an extra constraint to produce a smooth model by minimising the first-order spatial derivative of the solution estimate $\mathbf{x}$.

(2) The maximum entropy constraint is

$$R = \sum_i |x_i| \ln |x_i|.$$

As the standard L_2 norm model regularisation, where $\mathbf{x}_{\text{ref}} = \mathbf{0}$ is assumed, is $R = \|\mathbf{x}\|^2 = \sum_i x_i x_i$, its solution is biased to the strong amplitudes in $\mathbf{x}$ ('bright spot'). In the maximum entropy regularisation, the use of logarithmic values will reduce the influence of strong $\mathbf{x}$ values.

(3) The Cauchy distribution of the solution is

$$R = \sum_{i=1}^{N} \ln\left(\frac{1}{\pi} \frac{\lambda}{\lambda^2 + x_i^2}\right) = -N \ln(\pi\lambda) - \sum_{i=1}^{N} \ln\left(1 + \frac{x_i^2}{\lambda^2}\right),$$

where λ is the Cauchy parameter. The main purpose of the use of the Cauchy constraint is to produce a sparse solution estimate. ∎

21. In a linear inverse problem $\mathbf{Gx} = \mathbf{d}$, the least-squares solution may be represented as

$$\mathbf{x} = [\mathbf{G}^T\mathbf{G} + \mu\mathbf{I}]^{-1}\mathbf{G}^T\mathbf{d},$$

where $\mathbf{d}$ is a data vector, $\mathbf{x}$ is the model vector to be resolved, and $\mathbf{G}$ is an operator mapping from the model space to the data space.

Explain different physical meaning for μ, in terms of

(1) stabilisation in matrix inverse;

(2) pre-whitening in deconvolution;

(3) a trade-off parameter that balances the data matching and a model regularisation; and

(4) the ratio of data and model covariance matrices.

A: (1) For the problem $\phi(\mathbf{x}) = \| \mathbf{d} - \mathbf{Gx} \|^2$, the least-squares solution is obtained by setting $\partial\phi/\partial\mathbf{x} = -2\mathbf{G}^T(\mathbf{d} - \mathbf{Gx}) = 0$. That is,

$$\mathbf{G}^T\mathbf{Gx} = \mathbf{G}^T\mathbf{d}.$$

If the square matrix $[\mathbf{G}^T\mathbf{G}]$ is near singular, a small positive value μ is added to the diagonal elements, so that $[\mathbf{G}^T\mathbf{G} + \mu\mathbf{I}]$ is invertible, and the stabilised solution is

$$\mathbf{x} = [\mathbf{G}^T\mathbf{G} + \mu\mathbf{I}]^{-1}\mathbf{G}^T\mathbf{d}.$$

(2) In the convolutional model, a seismic trace $\{\tilde{d}_i\}$ is presented by

$$\tilde{d}_i = \sum_{k} w_k r_{i-k} + e_i,$$

where $\{w_k\}$ is the seismic wavelet, and $\{r_k\}$ is the reflectivity series. The goal of deconvolution is to recover the reflectivity series $\{r_k\}$ from the recorded seismic trace $\{\tilde{d}_i\}$, which contains data errors $\{e_i\}$. The objective is to minimise the misfit between the real observation and the synthetic trace in a least-squares sense:

$$\phi(\mathbf{x}) = \| \tilde{\mathbf{d}} - \mathbf{Gx} \|^2,$$

where $\tilde{\mathbf{d}}$ is the vector of data $\{\tilde{d}_i\}$, $\mathbf{G}$ is the wavelet matrix in which each column contains the wavelet $\{w_k\}$, properly padded with zeros in order to express discrete convolution, and $\mathbf{x}$ is the vector of the reflectivity series $\{r_k\}$. In the least-squares solution

$$\mathbf{x} = [\mathbf{G}^T\mathbf{G} + \mu\mathbf{I}]^{-1}\mathbf{G}^T\tilde{\mathbf{d}},$$

matrix $[\mathbf{G}^T\mathbf{G}]$ is equivalent to the autocorrelation of the seismic trace, and the small positive damping factor μ, used to stabilise the solution, is called a pre-whitening parameter. In practice, μ is taken as a small fraction of the maximum value of the diagonal elements in $[\mathbf{G}^T\mathbf{G}]$, and is often referred to as the percentage level of white noise in the data.

(3) In the least-squares inversion with a model constraint, the objective function is defined as

$$\phi(\mathbf{x}) = \parallel \mathbf{d} - \mathbf{Gx} \parallel^2 + \mu \parallel \mathbf{x} \parallel^2,$$

where μ is a trade off parameter that controls the contributions of data-fitting term and the model term in the solution.

Setting $\partial\phi/\partial\mathbf{x} = -2\mathbf{G}^T(\mathbf{d} - \mathbf{Gx}) + 2\mu\mathbf{x} = \mathbf{0}$, we obtain the least-squares solution as

$$\mathbf{x} = [\mathbf{G}^T\mathbf{G} + \mu\mathbf{I}]^{-1}\mathbf{G}^T\mathbf{d}.$$

(4) Considering the data and model covariance matrices, the objective function is defined as

$$\phi(\mathbf{x}) = (\mathbf{d} - \mathbf{Gx})^T\mathbf{C}_d^{-1}(\mathbf{d} - \mathbf{Gx}) + \mathbf{x}^T\mathbf{C}_x^{-1}\mathbf{x},$$

where $\mathbf{C}_d$ is the data covariance matrix, and $\mathbf{C}_x$ is the model covariance matrix. A least-squares solution is given by

$$\mathbf{x} = [\mathbf{G}^T\mathbf{C}_d^{-1}\mathbf{G} + \mathbf{C}_x^{-1}]^{-1}\mathbf{G}^T\mathbf{C}_d^{-1}\mathbf{d}.$$

If making an approximation,

$$\mu = \frac{\parallel \mathbf{C}_d \parallel}{\parallel \mathbf{C}_x \parallel},$$

this solution is exactly the same solution as the classical model-constrained least-squares inversion.   ∎

References

Aki K. and Richards P. G., 1980. *Quantitative Seismology*. W. H. Freeman & Co., San Francisco.

Backus G. E. and Gilbert F., 1968. The resolving power of gross Earth data. *Geophysical Journal of the Royal Astronomical Society* **16**, 169–205.

Backus G. E. and Gilbert F., 1970. Uniqueness in the inversion of inaccurate gross Earth data. *Philosophical Transactions of the Royal Society of London, Series A: Mathematical and Physical Sciences* **266**, 123–192.

Bérenger J. P., 1994. A perfectly matched layer for the absorption of electro-magnetic waves. *Journal of Computational Physics* **114**, 185–200.

Bérenger J. P., 1996. Three-dimensional perfectly matched layer for the absorption of electromagnetic waves. *Journal of Computational Physics* **127**, 363–379.

Berkhout A. J., 1984. *Seismic Resolution: Resolving Power of Acoustical Echo Techniques*. Geophysical Press, London.

Beylkin G., 1987. Discrete Radon transform. *IEEE Transactions on Acoustics, Speech and Signal Processing* **35**, 162–172.

Bishop T. N., Bube K. P., Cutler R. T., Langan R. T., Love P. L., Resnick J. R., Shuey R. T., Spindler D. A. and Wyld H. W., 1985. Tomographic determination of velocity and depth in laterally varying media. *Geophysics* **50**, 903–923.

Bleibinhaus F., Hole J. A., Ryberg T. and Fuis G. S., 2007. Structure of the California Coast Ranges and San Andreas Fault at SAFOD from seismic waveform inversion and reflection imaging. *Journal of Geophysical Research* **112**, B06315.

Bleibinhaus F. and Rondenay S., 2009. Effects of surface scattering in full waveform inversion. *Geophysics* **74**, WCC69–77.

Bleistein N., 1984. *Mathematical Methods for Wave Phenomena*. Academic Press, London.

Bleistein N., Cohen J. K., and Stockwell J. W., 2000. *Mathematics of Multi-dimensional Seismic Imaging, Migration and Inversion*. Springer-Verlag, New York.

Brenders A. J. and Pratt R. G., 2007. Efficient waveform tomography for lithospheric imaging: implications for realistic, two-dimensional acquisition geometries and low-frequency data. *Geophysical Journal International* **168**, 152–170.

Brossier R., Operto S. and Virieux J., 2010. Which data residual norm for robust elastic frequency-domain full waveform inversion? *Geophysics* **75**, R37–R46.

Seismic Inversion: Theory and Applications, First Edition. Yanghua Wang.
© 2017 John Wiley & Sons, Ltd. Published 2017 by John Wiley & Sons, Ltd.

Broyden C. G., 1967. Quasi-Newton methods and their application to function minimisation. *Mathematics of Computation* **21**, 368–381.

Broyden C. G., 1972. The convergence of a class of double-rank minimisation algorithms 1: general considerations. *IMA Journal of Applied Mathematics* **6**, 76–90.

Bunks C., Saleck F. M., Zaleski S. and Chavent G., 1995. Multiscale seismic waveform inversion. *Geophysics* **60**, 1457–1473.

Červený V., 2001. *Seismic Ray Theory*. Cambridge University Press, Cambridge, England.

Červený V., Klimeš L. and Pšenčík I., 2007. Seismic ray method: recent developments. *Advances in Geophysics* **48**, 1–126.

Claerbout J. F., 1992. *Earth Soundings Analysis: Processing versus Inversion*. Blackwell Scientific Publications, Boston.

Cohen L. and Lee C., 1989. Standard deviation of instantaneous frequency. *IEEE Proceedings of International Conference on Acoustics, Speech and Signal Processing* **4**, 2238–2241.

Connolly P., 1999. Elastic impedance. *The Leading Edge* **18**, 438–352.

Crase E., Pica A., Noble M., McDonald J. and Tarantola A., 1990. Robust elastic nonlinear waveform inversion: application to real data. *Geophysics* **55**, 527–538.

Debeye H. W. J. and van Riel P., 1990. L_p norm deconvolution. *Geophysical Prospecting* **38**, 381–403.

Deregowski S. M. and Brown S. M., 1983. A theory of acoustic diffractors applied to 2-D models. *Geophysical Prospecting* **31**, 293–333.

Farra V. and Madariaga R., 1988. Nonlinear reflection tomography. *Geophysical Journal* **95**, 135–147.

Fichtner A., 2010. *Full Seismic Waveform Modelling and Inversion*. Springer-Verlag, Berlin.

Fletcher R. and Reeves C. M., 1964. Function minimisation by conjugate gradients. *The Computer Journal* **7**, 149–154.

Fletcher R., 1976. Conjugate gradient methods for indefinite systems. In: Watson G. A. (ed.), *Numerical Analysis: Lecture Notes in Mathematics* **506**, pp.73–89. Springer-Verlag, New York.

Franklin J. N., 1970. Well-posed stochastic extension of ill-posed linear problems. *Journal of Mathematical Analysis and Applications* **31**, 682–716.

Gardner G. H. F., Gardner L. W. and Gregory A. R., 1974. Formation velocity and density: the diagnostic basics for stratigraphic traps. *Geophysics* **39**, 770–780.

Gauthier O., Virieux J. and Tarantola A., 1986. Two-dimensional nonlinear inversion of seismic waveforms: numerical results. *Geophysics* **51**, 1387–1403.

Golub G. and Kahan W., 1965. Calculating the singular values and pseudo-inverse of a matrix. *Journal of the Society for Industrial and Applied Mathematics Series B: Numerical Analysis* **2**, 205–224.

Golub G. and Reinsch C., 1970. Singular value decomposition and least squares solutions. *Numerische Mathematik* **14**, 403–420.

Graham R. L., Knuth D. E. and Patashnik O., 1994. *Concrete Mathematics: A Foundation for Computer Science*. Addison-Wesley Professional, Reading, Massachusetts.

Hadamard J., 1902. Sur les problèmes aux dérivées partielles et leur signification physique. *Princeton University Bulletin* **13**, 49–52.

Helmberger D. V., Song X. J. and Zhu L., 2001. Crustal complexity from regional waveform tomography: aftershocks of the 1992 Landers earthquake, California. *Journal of Geophysical Research* **106** (B1), 609–620.

Hestenes M. R. and Stiefel E., 1952. Methods of conjugate gradients for solving linear systems. *Journal of Research of the National Bureau of Standards* **49**, 409–436.

Hestholm S. O. and Ruud B. O., 1998. 3-D finite difference elastic wave modelling including surface topography. *Geophysics* **63,** 613–622.

Hestholm S. O. and Ruud B. O., 2000. 2-D finite-difference viscoelastic wave modelling including surface topography. *Geophysical Prospecting* **48**, 341–373.

Hoffman K. A. and Chiang S. T., 2000. *Computational Fluid Dynamics*. Engineering Education System, Wichita, Kansas.

Hole J. A. and Zelt B. C., 1995. 3-D finite-difference reflection travel times. *Geophysical Journal International* **121**, 427–434.

Hosken J. W. J., 1988. Ricker wavelets in their various guises. *First Break* **6**, 24–33.

Householder A. S., 1955. Terminating and nonterminating iterations for solving linear systems. *Journal of the Society for Industrial and Applied Mathematics* **3**(2), 67–72.

Householder A. S., 1964. *The Theory of Matrices in Numerical Analysis*. Blaisdell Publishing Co., New York.

Jaynes E. T., 1968. Prior probabilities. *IEEE Transactions on Systems Science and Cybernetics* **4**, 227–241.

Johnson N. L. and Kotz S., 1972. *Distributions in Statistics*: *Continuous Multivariate Distributions*. John Wiley & Sons, New York.

Julian B. R. and Gubbins D., 1977. Three-dimensional seismic ray tracing. *Journal of Geophysics* **43**, 95–113.

Käser M. and Igel H., 2001. Numerical simulation of 2-D wave propagation on unstructured grids using explicit differential operators. *Geophysical Prospecting* **49**, 607–619.

Kennett B. L. N., Sambridge M. S. and Williamson P. R., 1988. Subspace methods for large inverse problems with multiple parameter classes. *Geophysical Journal* **94**, 237–247.

Kim S., 2002. 3D eikonal solvers: first-arrival traveltimes. *Geophysics* **67**, 1225–1231.

Komatitsch D., Coute F. and Mora P., 1996. Tensorial formulation of the wave equation for modelling curved interfaces. *Geophysical Journal International* **127**, 156–168.

Kosloff D. D. and Sudman Y., 2002. Uncertainty in determining interval velocities from surface reflection seismic data. *Geophysics* **67**, 952–963.

Lanczos C., 1950. An iteration method for the solution of the eigenvalue problem of linear differential and integral operators. *Journal of Research of the National Bureau of Standards* **45**, 255–282.

Levy S. and Fullagar P. K., 1981. Reconstruction of a sparse spike train from a portion of its spectrum and application to high-resolution deconvolution. *Geophysics* **46**, 1235–1243.

Lombard B., Piraux J., Gelis C. and Virieux J., 2008. Free and smooth boundaries in 2-D finite-difference schemes for transient elastic waves. *Geophysical Journal International* **172**, 252–261.

Longbottom J., Walden A. T. and White R. E., 1988. Principles and application of maximum kurtosis phase estimation. *Geophysical Prospecting* **36**, 115–138.

Mendel J. M., 1991. Tutorial on higher-order statistics (spectra) in signal processing and system theory: theoretical results and some applications. *Proceedings of the IEEE* **79**, 278–305.

Moore E. H., 1920. On the reciprocal of the general algebraic matrix. *Bulletin of the American Mathematical Society* **26**, 394–395.

Mora P., 1987. Nonlinear two-dimensional elastic inversion of multi-offset seismic data. *Geophysics* **52**, 1211–1228.

Mora P., 1988. Elastic wavefield inversion of reflection and transmission data. *Geophysics* **53**, 750–759.

Morgan J., Warner M., Bell R., Ashley J., Barnes D., Little R., Roele K. and Jones C., 2013. Next-generation seismic experiments: wide-angle, multi-azimuth, three-dimensional, full-waveform inversion. *Geophysical Journal International* **195**, 1657–1678.

Nocedal J. and Wright S. J., 2006. *Numerical Optimisation*. Springer-Verlag, Berlin.

Nolet G. (ed.), 1987. *Seismic Tomography: with Applications in Global Seismology and Exploration Geophysics*. D. Reidel, Dordrecht, Holland.

Operto S., Virieux J., Dessa J.-X. and Pascal G., 2006. Crustal seismic imaging from multifold ocean bottom seismometer data by frequency domain full waveform tomography: application to the eastern Nankai trough. *Journal of Geophysical Research* **111**, B09306.

Patantonis D. E. and Atharassiadis N. A., 1985. A numerical procedure for the generation of orthogonal body-fitted coordinate systems with direct determination of grid points on the boundary. *International Journal for Numerical Methods in Fluids* **5**, 245–255.

Penrose R., 1955. A generalised inverse for matrices. *Proceedings of the Cambridge Philosophical Society* **51**, 406–413.

Plessix R.-E. and Li Y., 2013. Waveform acoustic impedance inversion with spectral shaping. *Geophysical Journal International* **195**, 301–314.

Pollitz F. F. and Fletcher J. P., 2005. Waveform tomography of crustal structure in the south San Francisco Bay region. *Journal of Geophysical Research* **110**, B08308.

Pratt R. G. and Worthington M. H., 1990. Inverse theory applied to multisource crosshole tomography, part I: acoustic wave-equation method. *Geophysical Prospecting* **38**, 287–310.

Pratt, R. G., Song Z. M., Williamson P. and Warner M., 1996. Two-dimensional velocity models from wide-angle seismic data by wavefield inversion. *Geophysical Journal International* **124**, 323–340.

Pratt R. G., 1999. Seismic waveform inversion in the frequency domain, Part 1: theory and verification in a physical scale model. *Geophysics* **64**, 888–901.

Priestley K., Debayle E., McKenzie D. and Pilidou S., 2006. Upper mantle structure of eastern Asia from multimode surface waveform tomography. *Journal of Geophysical Research* **111**, B10304.

Qin F., Luo Y., Olsen K. B., Cai W. and Schuster G. T., 1992. Finite-difference solution of the eikonal equation along expanding wavefronts. *Geophysics* **57**, 478–487.

Rao Y. and Wang Y., 2013. Seismic waveform simulation with pseudo-orthogonal grids for irregular topographic models. *Geophysical Journal International* **194**, 1778–1788.

Rao Y., Wang Y., Chen S. and Wang J., 2016. Crosshole seismic tomography with cross-firing geometry. *Geophysics* **81**(4), R139–R146.

Ravaut C., Operto S., Improta L., Virieux J., Herrero A. and Dell'Aversana P., 2004. Multiscale imaging of complex structures from multifold wide-aperture seismic data by frequency-domain full-waveform tomography: application to a thrust belt. *Geophysical Journal International* **159**, 1032–1056.

Rawlinson N. and Sambridge M., 2003. Seismic traveltime tomography of the crust and lithosphere. *Advances in Geophysics* **46**, 81–197.

Rawlinson N., Hauser J. and Sambridge M., 2007. Seismic ray tracing and wavefront tracking in laterally heterogeneous media. *Advances in Geophysics* **49**, 203–273.

Rawlinson N., Pozgay S. and Fishwick S., 2010. Seismic tomography: a window into deep Earth. *Physics of the Earth and Planetary Interiors* **178**, 101–135.

Rawlinson N., Fichtner A., Sambridge M. and Young M., 2014. Seismic tomography and the assessment of uncertainty. *Advances in Geophysics* **55**, 1–76.

Ricker N., 1953. The form and laws of propagation of seismic wavelets. *Geophysics* **18**, 10–40.

Robinson E. A. and Treitel S., 1980. *Geophysical Signal Analysis*. Prentice Hall, New Jersey.

Santosa F. and Schwetlick H., 1982. The inversion of acoustical impedance profile by methods of characteristics. *Wave Motion* **4**, 99–110.

Sheriff R. E., 1991. *Encyclopaedic Dictionary of Exploration Geophysics*. Society of Exploration Geophysicists, Tulsa.

Snieder R., Xie M. Y., Pica A. and Tarantola A., 1989. Retrieving both the impedance contrast and background velocity: a global strategy for the seismic reflection problem. *Geophysics* **54**, 991–1000.

Tarantola A., 1984. Inversion of seismic reflection data in the acoustic approximation. *Geophysics* **49**, 1259–1266.

Tarantola A., 1986. A strategy for nonlinear elastic inversion of seismic reflection data. *Geophysics* **51**, 1893–1903.

Tarantola A., 1987. *Inverse Problem Theory: Methods for Data Fitting and Parameter Estimation*. Elsevier Science, Amsterdam.

Tarrass I., Giraud L. and Thore P., 2011. New curvilinear scheme for elastic wave propagationg in presence of curved topography. *Geophysical Prospecting* **59**, 889–906.

Thomas P. D. and Middlecoeff J. F., 1980. Direct control of the grid point distribution in meshes generated by elliptic equations. *AIAA Journal* **18**, 652–656.

Thompson J. F., Warsi Z. U. A. and Mastin C. W., 1985. *Numerical Grid Generation: Foundations and Applications*. North-Holland, Amsterdam.

Tikhonov A. N., 1935. Theorèmes d'unicité pour l'équation de la chaleur. *Matematiceskij Sbornik* **42**, 199–216.

Tikhonov A. N. and Arsenin V. Y., 1977. *Solutions of Ill-Posed Problems*. John Wiley & Sons Inc., New York.

Tikhonov A. N., Goncharsky A. V., Stepanov V. V. and Yagoda A. G., 1995. *Numerical Methods for the Solution of Ill-Posed Problems*. Springer, Dordrecht.

van der Vorst H. A., 1992. Bi-CGStab: A fast and smoothly converging variant of BiCG for the solution of non-symmetric linear systems. *SIAM Journal on Scientific and Statistical Computing* **13**, 631–644.

van Trier J. and Symes W. W., 1991. Upwind finite-difference calculation of traveltimes. *Geophysics* **56**, 812–821.

Vidale J. E., 1990. Finite-difference calculations of traveltimes in three dimensions. *Geophysics* **55**, 521–526.

Vigh D., Starr E. and Kapoor J., 2012. Developing Earth models with full waveform inversion. *The Leading Edge* **28**, 432–435.

Virieux J., Calandra H. and Plessix R. E., 2011. A review of the spectral, pseudo-spectral, finite-difference and finite-element modelling techniques for geophysical imaging. *Geophysical Prospecting* **59**, 794–813.

Wang Y. and Houseman G. A., 1994. Inversion of reflection seismic amplitude data for interface geometry. *Geophysical Journal International* **117**, 92–110.

Wang Y. and Houseman G. A., 1995. Tomographic inversion of reflection seismic amplitude data for velocity variation. *Geophysical Journal International* **123**, 355–372.

Wang Y. and Pratt R. G., 1997. Sensitivities of seismic traveltimes and amplitudes in reflection tomography. *Geophysical Journal International* **131**, 618–642.

Wang Y., 1999a. Approximations to the Zoeppritz equations and their use in AVO analysis. *Geophysics* **64**, 1920–1927.

Wang Y., 1999b. Random noise attenuation using forward-backward linear prediction. *Journal of Seismic Exploration* **8**, 133–142.

Wang Y., 1999c, Simultaneous inversion for model geometry and elastic parameters. *Geophysics* **64**, 182–190.

Wang Y. and Pratt R. G., 2000. Seismic amplitude inversion for interface geometry of multi-layered structures. *Pure and Applied Geophysics* **157**, 1601–1620.

Wang Y., 2002. Seismic trace interpolation in the *f-x-y* domain. *Geophysics* **67**, 1232–1239.

Wang Y., 2003a. Multiple attenuation: coping with the spatial truncation effect in the Radon transform domain. *Geophysical Prospecting* **51**, 75–87.

Wang Y., 2003b. *Seismic Amplitude Inversion in Reflection Tomography*. Pergamon, Elsevier Science, Amsterdam.

Wang Y., 2004a. *Q* analysis on reflection seismic data. *Geophysical Research Letters* **31**, L17606.

Wang, Y., 2004b. Multiple prediction through inversion: a fully data-driven concept for surface-related multiple attenuation. *Geophysics* **69**, 547–553.

Wang Y., 2006. Inverse *Q*-filter for seismic resolution enhancement. *Geophysics* **71**, V51–V60.

Wang Y. and Rao R., 2006. Crosshole seismic waveform tomography I: strategy for real data application. *Geophysical Journal International* **166**, 1237–1248.

Wang Y., 2007. Multiple prediction through inversion: theoretical advancements and real data application. *Geophysics* **72**, V33–V39.

Wang Y. and Rao Y., 2009. Reflection seismic waveform tomography. *Journal of Geophysical Research* **114**, B03304.

Wang Y., 2013. Simultaneous computation of seismic slowness paths and the traveltime field in anisotropic media. *Geophysical Journal International* **195**, 1141–1148.

Wang Y., 2014. Seismic ray tracing in anisotropic media: a modified Newton algorithm for solving highly nonlinear systems. *Geophysics* **79**, T1–T7.

Wang Y., 2015a. Frequencies of the Ricker wavelet. *Geophysics* **80**, A31–A37.

Wang Y., 2015b. Generalised seismic wavelets. *Geophysical Journal International* **203**, 1172–1178.

Wang Y., 2015c. The Ricker wavelet and the Lambert *W* function. *Geophysical Journal International* **200**, 111–115.

Warner M., Ratcliffe A., Nangoo T., Morgan J., Umpleby A., Shah N., Vinje V., Stekl I., Guasch L., Win C., Conroy G. and Bertrand A., 2013. Anisotropic 3-D full-waveform inversion. *Geophysics* **78**, R59–R80.

White R. E., 1984. Signal and noise estimation from seismic reflection data using spectral coherence methods. *Proceedings of the IEEE* **72**, 1340–1356.

White R. E., 1988. Maximum kurtosis phase correction. *Geophysical Journal International* **95**, 371–389.

White R. E., Simm R. and Xu S. Y., 1998. Well tie, fluid substitution and AVO modelling: a North Sea example. *Geophysical Prospecting* **46**, 323–336.

White R. E. and Simm R., 2003. Tutorial: good practice in well ties. *First Break* **21**, 75–83.

Yang W. C., 1997. *Theory and Methods of Geophysical Inversion* (in Chinese). Geological Publishing House, Beijing.

Zelt C. A., 2011. Traveltime tomography using controlled-source seismic data. In: Gupta H. K. (ed.), *Encyclopedia of Solid Earth Geophysics*, pp.1453–1473. Springer-Verlag, Berlin.

Zhang J. F. and Liu T. L., 1999. P-SV-wave propagation in heterogeneous media: grid method. *Geophysical Journal International* **136**, 431–438.

Zhang J. F., 2004. Wave propagation across fluid-solid interfaces: a grid method approach. *Geophysical Journal International* **159**, 240–252.

Zhang W. and Chen X. F., 2006. Traction image method for irregular free surface boundaries in finite-difference seismic wave simulation. *Geophysical Journal International* **167**, 337–353.

Author index

Seismic Inversion: Theory and Applications, First Edition. Yanghua Wang.
© 2017 John Wiley & Sons, Ltd. Published 2017 by John Wiley & Sons, Ltd.

Subject index

Seismic Inversion: Theory and Applications, First Edition. Yanghua Wang.
© 2017 John Wiley & Sons, Ltd. Published 2017 by John Wiley & Sons, Ltd.

Printed and bound by CPI Group (UK) Ltd, Croydon, CR0 4YY

27/10/2024

14580301-0001